TEST PREP MATH BOOK
FOR CASAS MATH GOALS 2 LEVEL E

Helping Learners Approach Math with Confidence while Preparing them for CASAS Math GOALS 2 Level E—**Forms 929M and 930M**

By

CBL | COACHING FOR BETTER LEARNING

PREFACE

Dear Instructors,

This Test Prep math book is specifically designed to prepare adult learners for the CASAS Math GOALS 2 Level E Forms 929M and 930M. It fully aligns with the CASAS Competencies and meets the requirements of the College and Career Readiness Standards (CCRS), the National Reporting System (NRS), and the Workforce Innovation and Opportunity Act (WIOA).

Adhering to the CASAS test blueprint, this textbook covers mathematical areas through six detailed chapters: *Number Sense and Operations, Consumer Economics, Algebraic Thinking, Geometry and Measurement, Data Analysis, Statistics and Probability, and Pure Mathematics.*

The book's content is structured to improve the mathematical thinking skills of adult students. It provides 14 lessons across various competencies, such as Consumer Economics, Community Resources, Employment, and Pure Mathematics. Each chapter is thoughtfully crafted with multiple lessons to foster deep understanding and practical application of mathematical concepts.

Additionally, the book includes two practice tests that simulate the actual CASAS assessments, incorporating real-world math problems to give students a genuine taste of what they can expect. Complete with answer keys for all exercises and practice tests, this textbook is a robust tool for effective learning and assessment.

Using this resource in your teaching will equip you to effectively develop and improve the math strategies, functions, and concepts necessary for your adult learners' success. Indeed, this book is an invaluable asset for programs that aim to empower their students with the mathematical skills required for success in everyday contexts such as community involvement, family management, and professional environments. To order class sets, go to cbledu.com.

INTRODUCTION

Dear Math Students,

Welcome to your journey toward improving your math skills with this test-prep math textbook. It is designed specifically for adult learners like you. This book is structured to prepare you for the CASAS Math GOALS 2 Level E test. In the six chapters, you'll explore essential mathematical areas, including *Number Sense and Operations, Consumer Economics, Algebraic Thinking, Geometry and Measurement, Data Analysis, Statistics and Probability, and Pure Mathematics.* With 14 practical lessons, this textbook offers a clear path to improving your mathematical knowledge.

Practicing the exercises in this book is crucial. Each chapter includes multiple lessons that build on each other to help you understand and apply mathematical concepts in real-world situations. By engaging with these exercises, you'll develop a stronger foundation in each topic, making sure that you're well-prepared not just for the tests but also for the practical application of these skills in daily life.

We've also included two practice tests that mimic the actual CASAS level E test. These practice tests are designed to give you a realistic experience of what to expect on the actual test. By taking these practice tests, you can assess your progress, identify areas where you need further practice, and build your confidence.

This textbook is more than just a study guide—it's a tool that will equip you with the math strategies, functions, and concepts necessary to solve word problems confidently. Regular practice and study will transform your understanding of math, turning challenges into opportunities for growth and learning.

Remember, math skills are essential for success in various aspects of your life, including community involvement, managing family finances, and professional advancement. By committing to completing the exercises and fully engaging with the materials in this book, you'll be setting yourself up for success in your academic pursuits and beyond.

Let's get started on this path together!

HOW TO APPROACH MATH

Here are ten practical ways you can overcome math fear and anxiety and build confidence while using this math textbook:

1. **Start Small:** Begin with easier problems that you can solve to build your confidence before solving harder ones.
2. **Practice Regularly:** Consistent practice makes math feel more manageable. Try to work on math problems a few times a week.
3. **Use the Book's Resources:** Take advantage of the tools and explanations in your textbook. They are designed to help you understand and solve math problems.
4. **Take Breaks:** If you feel overwhelmed, take a short break. Come back to the problem with a clear mind.
5. **Ask for Help:** Don't hesitate to seek help when you need it. Ask a teacher, a classmate, or use online resources if you're stuck.
6. **Stay Positive:** Keep a positive attitude about math. Remind yourself that you can handle it and that it's okay to make mistakes as you learn.
7. **Set Small Goals:** Break your math studies into small, achievable goals. Celebrate when you reach these goals to motivate yourself.
8. **Understand, Don't Memorize:** Focus on understanding the math concepts rather than just memorizing formulas. This understanding will make you feel more confident in your ability to solve math problems.
9. **Visualize Success:** Picture yourself successfully solving problems and understanding concepts. This visualization can boost your confidence.
10. **Reflect on Progress:** Regularly look back at where you started and recognize the progress you've made. This can be a great confidence booster.

By following these strategies, you'll be better positioned to tackle math with less anxiety and more confidence.

STUDY STRATEGIES

Here are ten simple strategies to study and improve your math knowledge, skills, and understanding using this textbook. Each strategy is designed to be practical and straightforward.

Strategy	Description
1. Set a study schedule.	Allocate specific times each week for math study and practice to build a routine.
2. Create a study space.	Find a quiet, organized space dedicated to studying to stay focused.
3. Use the textbook.	Read explanations and solve problems in the textbook to understand concepts deeply.
4. Practice with examples.	Work through example problems to understand how to apply math rules before trying exercises on your own.
5. Summarize each lesson.	Write a brief summary of what you learned in each lesson to reinforce your understanding.
6. Solve practice tests.	Use practice tests in the textbook to prepare for the actual test and build confidence.
7. Discuss with peers.	Study in groups or discuss problems with classmates to get different perspectives and solutions.
8. Teach someone else.	Explain math concepts to someone else to improve your own understanding and retention.
9. Use online resources.	Supplement your textbook with online tutorials and exercises (e.g., YouTube videos) for additional practice.
10. Review regularly.	Regularly go back and review previous chapters to keep information fresh and build connections between topics and chapters.

Using these strategies can help you make the most of your math textbook and your study time to gradually build up your mathematical abilities and confidence.

CHAPTER 1
NUMBER SENSE AND OPERATIONS

Lesson 1: Solving Multi-Step Problems Using Rates and Proportional Relationships

A **ratio** is a comparison of two numbers or quantities. They are measured in the same or similar units. **Rate** is a type of ratio that compares two quantities measured in **different types of units.**

Example 1:

If it takes Megan 2 hours to drive 20 miles, how many hours will it take her to drive 45 miles at the same rate?

Solution:

If Megan drives 20 miles in 2 hours, her rate is:

$$rate = \frac{20\ miles}{2\ hours} = 10\ miles\ per\ hour$$

Substitute the previous rate into the distance equation and solve for the time:

$$distance = rate\ \times time$$

$$time = \frac{distance}{rate}$$

$$\Rightarrow\ time = \frac{45\ miles}{10\ miles\ per\ hour} = 4.5\ hours$$

Proportional relationships are relationships between two variables where **their ratios are equivalent.** An equation of two equivalent ratios is called **proportion**.

Example 2:

A 5-pound bag of candy costs $6. How much does each pound of candy cost?

Solution:

We can set up the following proportion that represents the problem:

$$\frac{5\ pound}{\$6} = \frac{1\ pound}{x}$$

Solve for x:

$$\frac{5\ pound}{\$6} = \frac{1\ pound}{x}$$

$$(5\ pound) \cdot x = (1\ pound) \cdot \$6$$

$$x = \frac{(1\ pound) \cdot \$6}{5\ pound} = \$1.2$$

Then, each pound of candy costs $1.2.

Practice Exercises

$$\frac{2\ gallons}{3\ minutes} = \frac{x}{24\ minutes}$$

1. What is *x* in this proportion?

   A. 16 minutes
   C. 16 gallons
   B. 36 minutes
   D. 36 gallons

2. A dog runs at a rate of 15 miles per hour. How far does he run in 20 minutes?

   A. 5 miles
   C. 20 miles
   B. 45 miles
   D. 1.5 miles

3. In a party, 18 drinks are required for every 15 people. If there are 240 people, how many soft drinks are required?

   A. 320
   C. 250
   B. 200
   D. 288

The price of 3 T-shirts at Super Market is $27. The price of 4 of the same T-shirts at John's Store is $34. At Walkmart, the price of a dozen of the same T-shirts is $117.60.

4. What is the cost of each T-shirt at Super Market?

   A. $8.00
   B. $9.00
   C. $9.50
   D. $7.50

5. What is the cost of each T-shirt at Walkmart?

   A. $9.80
   B. $8.50
   C. $9.50
   D. $9.75

6. Which place offers the best price?

   A. Walkmart
   B. Super Market
   C. John's Store
   D. None of the above

7. What is the cost of 10 T-shirts at Walkmart?

   A. $100
   B. $95
   C. $110
   D. $98

8. The bakers at a bakery can make 105 croissants in 3 hours and 30 minutes. What is the rate per hour?

   A. 32 croissants per hour
   B. 35 croissants per hour
   C. 30 croissants per hour
   D. 37 croissants per hour

9. Justin has saved $180 in the past 15 days. At this rate, how much will he save in six weeks?

   A. $72
   B. $120
   C. $365
   D. $504

10. What is the cost of 50 eggs?

A. $5.79

B. $2.99

C. $7.89

D. $6.00

Answer Keys:

1) C
2) A
3) D
4) B
5) A
6) C
7) D
8) C
9) D
10) A

Lesson 2: Understanding Radicals and Irrational Numbers

An **irrational number** is a number that cannot be written as a fraction. Whenever a number is preceded with a radical sign (√), the number is called a **radical.**

Irrational numbers

$$\sqrt{2}, \sqrt{5}, \sqrt[3]{7}, \pi$$

Not all radicals are irrational. For example, $\sqrt{9}$ is not an irrational number. It is because $\sqrt{9} = 3$ and 3 is a whole number.

Example: Is $\sqrt[3]{30}$ an irrational number?

Solution:

Using a calculator we get:

$$\sqrt[3]{30} = 3.10732\ldots$$

Notice that the decimal expansion is non terminating and non-repeating, so $\sqrt[3]{30}$ is irrational.

Practice Exercises

1. Which of the following is true?

   A. Pete said that all square roots are irrational numbers.

   B. 0.44 is an irrational number.

   C. $\sqrt{25}$ is not an irrational number.

   D. Tamara said that all radicals are irrational numbers.

Look at the following formula:
$m = \sqrt{6} \cdot k$

2. Let k = 2. Which of the following is true?

   A. m is a whole number.

   B. $m = \sqrt{2}$

   C. m is a radical.

   D. $m = \sqrt{62}$

3. Let k = 6. Which of the following is true?

   A. m is an irrational.

   B. m = 18

   C. $m = \sqrt{12}$

   D. m = 6

4. Which value of k makes m = 12?

   A. 2

   B. 12

   C. 4

   D. 24

5. Which of the following is an irrational number?

   A. 1.345

   B. $\sqrt{81}$

   C. $1 + \sqrt{1}$

   D. $\sqrt{13}$

Given:

$M = \sqrt{2}$ $\quad N = \sqrt{16}$

$P = \sqrt{3}$ $\quad Q = \sqrt{64}$

6. Which expression results in an irrational number?

   A. N + 1

   B. M + P

   C. N + Q

   D. Q – N

7. Which of the following is false?

   A. Q is greater than 10.

   B. N is equal to 4.

   C. M is lower than P.

   D. N + Q = 12

8. Which expression is greater than 5?

A. M + P

B. Q – N

C. N – P

D. M + N

9. The side length of a square is the square root of the area. The area of a square is 121 square inches. Which is the side length of the square?

A. 10 inches

B. 11 inches

C. 60.5 inches

D. $\sqrt{11}$ inches

10. Nancy estimated the distance across a pond as $\sqrt{460}$ feet. Which of the following represents the approximate distance across the pond? (Round your answer to the nearest hundredth.)

A. 21.50 ft.

B. 21.45 ft.

C. 22.44 ft.

D. 20.47 ft.

Answer Keys:

1) C
2) C
3) D
4) D
5) D
6) B
7) A
8) D
9) B
10) B

REFLECTION ON LEARNING

Answer the following reflection questions and feel free to discuss your responses with your teacher or a classmate.

1- What math ideas and principles did you learn in this chapter?

2- What new math concepts did you learn?

3- What procedures or methods did you practice in this chapter?

4- What aspect of this chapter is still not 100% clear for you?

5- What else do you want your teacher to know?

CHAPTER 2
MATH FOR DAILY TRANSACTIONS

Lesson 1: Measurement and Money

Money is a medium of exchange that comes in different forms, like coins or notes, that the government issues or publishes. People use money every day. Being able to count it and calculate how much change we should receive during a transaction is an important life skill.

Currency is the money system or the standard form of money used in a country for general use. It is the money people of a particular nation use. Here are some examples of currencies: U.S. dollars (US $), Euros (€) and Japanese Yen (¥).

Example 1:

Express 30 dollar in dimes.

Solution:

We know that 1 dollar = 10 dimes.
Then, 30 dollars = 30 x 10 = 300 dimes

Example 2:

Joel wants to buy a soccer ball from an online German site and he is in the USA. The cost of the ball is €60. How much does the ball cost in dollars?

Solution:

We know that 1 euro is equal to 1.08 USD. Then, the value of 60 Euro (€) is equal to:

$$60 \times 1.08 = 64.80$$

The soccer ball cost is $64.80.

Practice Exercises

1. Betty has 62 dollar bills and 9 quarters. How much money does she have?

   A. $62

   B. $62.50

   C. $64.25

   D. $63

> 1 euro = 1.08 dollar
> 1 dollar = 82.88 rupees

2. Convert 50 euros to dollars.

   A. $62

   B. $54

   C. $58

   D. $60

3. Convert 300 rupees to dollars.

   A. $3.62

   B. $0.30

   C. $1.53

   D. $5.00

4. Convert 100 euros to rupees.

   A. 1.30 rupees

   B. 8,288 rupees

   C. 7,674.07 rupees

   D. 8,951.04 rupees

Mr. Woods is buying some items from an online shop. He buys a smartphone costing $420, a chair costing $85, and a jacket costing $112.

1 dollar = 0.92 euro

1 euro = 162.61 Yen

5. What is the cost of the smartphone in euros?

   A. 386.40 euros
   B. 456.52 euros
   C. 433.50 euros
   D. 120.96 euros

6. What is the cost of the jacket in Yens?

   A. 19,796 Yens
   B. 18,144 Yens
   C. 17,235.98 Yens
   D. 16,755.33 Yens

7. What is the cost of his bill in euros?

   A. 617 euros
   B. 670.65 euros
   C. 715.66 euros
   D. 567.64 euros

8. What is the value of 50 dollars in nickels? (Hint: 1 Nickel = 5 cents)

   A. 250 nickels
   B. 500 nickels
   C. 750 nickels
   D. 1,000 nickels

9. If 1 dollar = 150.23 Yens, 1 Yen is equivalent to:

   A. 0.66 dollar
   B. 0.0067 dollar
   C. 0.066 dollar
   D. 0.70 dollar

10. Julia makes $452.50 each month. How much money will she make in 2 years and 6 months?

A. $13,575

B. $14,111.76

C. $10,860

D. $43,440

Answer Keys:

1) C
2) B
3) A
4) D
5) A
6) D
7) D
8) D
9) B
10) A

Lesson 2: Math for Purchasing Goods and Services

A **good** is a physical item that can be bought, touched, and used. A **service** is an action done for people who pay for it.

A **discount** is a percentage or dollar amount that represents how much someone doesn't have to pay. For example, a discount of 10% means we do not have to pay 10% of the original price. Similarly, a discount of \$8 means we pay \$8 less than the original price.

To compare prices, divide the cost by the weight or quantity of the item. Then, we can compare the two items. In other words, to compare prices, we compare the **unit rates** of the items.

Example 1:

The price of a bike is \$480. How much do you need to pay if you get a 25% discount?

Solution:

25% of \$480 = 0.25 x \$480 = \$120

We need to pay \$480 – \$120 = \$360

Example 2:

At Joshua's Store, Mike could get 3 PlayStation 5 consoles for \$1,350. At an online shop, the price for 4 of these is \$1,860. Which place has the better price for a PlayStation 5 console?

Solution:

Find the unit price for each place:

Joshua's Store:

$$unit\ rate = \frac{\$1{,}350}{3} = \$450\ per\ console$$

Online Store:

$$unit\ rate = \frac{\$1{,}860}{4} = \$465\ per\ console$$

Therefore, Joshua's Store offers the better price for a PlayStation 5 console (\$450)

Practice Exercises

Look at the following gas bill.

▶Account Summary

Previous Amount Due on 05/05/2022	$87.99
Payments Received by 05/11/2022 Thank You	$0.00
Balance on 05/11/2022	$87.99
Charges for Gas Service This Period	+$31.23
Please Pay This Amount	**$119.22**
Amount Past Due - Pay Immediately	**$87.99**
Current Charges Due by 06/06/2022	**$31.23**

- For more information regarding these charges, see the Detail Charges section.

1. What is the total amount due?

   A. $87.99 C. $31.23

   B. $119.22 D. $56.76

2. If the gas bill increased by 8% the next month, what would be the total amount due?

   A. $127.22 C. $130.55

   B. $128.76 D. $109.68

3. The gas bill was 15% less than the previous month. What was the total amount due?

   A. $104.22 C. $100.45

   B. $137.10 D. $101.34

4. Janet and her mother are eating out for dinner. The bill was $56.80. She wants to leave a 12% tip. How much should she leave for a tip?

   A. $6.82 C. $7.15

   B. $12.00 D. $5.83

Look at the following grocery receipt:

GROCERY DEPOT
5000 GA-5
Douglasville, GA 30135

Cashier: ENZO G.

DELITE SKIM	$10.36 TFA
4EA @ 2.59/EA	
WHOLEMILK	$7.77 TFA
3EA @ 2.59/EA	
REDBULL	$1.89 TFA
STRING CHEESE 16PK	$7.98 TFA
2EA @ 3.99/EA	
SUBTOTAL	$28.00
TAX	$1.82
TOTAL	$29.82
TEND	$29.82
CHANGE DUE	$0.00

Item Count 10

Thanks!!!

Date	Time	Lane	Clerk	Trans#
01/07/2019	09:45 AM	4	101	3854

5. How many items were purchased?

   A. 4

   B. 8

   C. 10

   D. 7

6. Which item was the most expensive?

   A. Delite Skim

   B. String cheese

   C. Wholemilk

   D. Redbull

7. Which item was the cheapest?

   A. Redbull

   B. Delite Skim

   C. String cheese

   D. Wholemilk

8. Suppose the tax was 8%. What would have been the total amount?

   A. $37.22

   B. $36.00

   C. $38.35

   D. $30.24

Coffee pods

Box of 24: **$19.99**
.83 cents each

Box of 48: **$67.00**
1.40 cents each

Box of 96: **$49.52**
.52 cents each

9. What is the best deal?

   A. Box of 96 coffee pods

   B. Box of 24 coffee pods

   C. Box of 48 coffee pods

10. What is the total cost of three boxes of 24 coffee pods and two boxes of 96 coffee pods?

    A. $139.02

    B. $159.01

    C. $193.97

    D. $233.04

Answer Keys:

1) B
2) B
3) D
4) A
5) C
6) B
7) A
8) D
9) A
10) B

REFLECTION ON LEARNING

Answer the following reflection questions and feel free to discuss your responses with your teacher or a classmate.

1- What math ideas and principles did you learn in this chapter?

2- What new math concepts did you learn?

3- What procedures or methods did you practice in this chapter?

4- What aspect of this chapter is still not 100% clear for you?

5- What else do you want your teacher to know?

CHAPTER 3
ALGEBRAIC THINKING

Lesson 1: Solving Problems Involving Inequalities, Pairs of Simultaneous Linear Equations and Quadratic Expressions

Inequalities are mathematical expressions involving the symbols >, <, ≥ and ≤. Solving an inequality means finding the range of values that make the inequality true for an unknown value (x).

Symbols

> means "is greater than"

≥ means "is greater than or equal to"

< means "is less than"

≤ means "is less than or equal to"

Inequality Key Words

at least - *means "greater than or equal to"*

no more than - *means "less than or equal to"*

more than - *means "greater than"*

at most - means "is less than or equal to"

Example:

Inequality

$$2x - 6 \leq 8$$

Simultaneous linear equations are two or more algebraic equations that share variables such as x and y. They are called simultaneous equations because the equations are solved at the same time.

$$3x - y = 5$$
$$2x + 5y = 9$$

A quadratic expression is an algebraic expression with the variable with the highest power of 2.

Quadratic Expressions

$$\boldsymbol{x^2 + 3}$$
$$\boldsymbol{2a^2 - b}$$

Example 1:

Robert has $900 in a savings account at the beginning of the winter. He wants to have at least $450 in the account by the end of the winter. He withdraws $75 each week for food, bills, and entertainment. How many weeks can Robert withdraw money from his account?

Solution:

Let x be the number of weeks it can Robert withdraw money. Notice that the word **at least** means the problem must be written as an inequality with the sign ≥.

We write the following inequality that represents the problem:

$$900 - 75x \geq 450$$

Now, solve the inequality.

$$900 - 75x \geq 450$$

$$-75x \geq 450 - 900$$

$$-75x \geq -450$$

$$\frac{-75x}{-75} \geq \frac{-450}{-75}$$

$$x \leq 6$$

Recall: Reverse the sign of the inequality when multiplying or dividing by a negative number.

Then, the number of weeks that Robert can withdraw money from his account is **6 weeks or less.**

Example 2:

In a school festival, 70 tickets were sold. Adult tickets cost \$6.00, children's cost \$3.00, and a total of \$270 was collected. How many tickets of each kind were sold?

Solution:

Step 1: Let x be the number of adult tickets. Let y be the number of children's tickets. We write two equations that represent the problem:

1) Total number of tickets: $x + y = 70$
2) Total money collected: $6x + 3y = 270$

So, we got the simultaneous linear equations that represent the problem:

$$1)\ x + y = 70$$

$$2)\ 6x + 3y = 270$$

Step 2: Solve the system of linear equations. Isolate y in the first equation:

$$x + y = 70 \quad \Rightarrow \quad y = 70 - x$$

Step 3: Replace this value of y into the second equation:

$$6x + 3(70 - x) = 270$$

Step 4: Solve the equation.

$$6x + 3(70 - x) = 270$$

$$6x + 210 - 3x = 270$$

$$6x - 3x = 270 - 210$$

$$3x = 60$$

$$x = \frac{60}{3} = 20$$

Step 5: Now, put the value of x in the first equation.

$$x + y = 70$$

$$20 + y = 70$$

$$y = 70 - 20 = 50$$

Then, the number of adult tickets is 20, and the number of children's tickets is 50.

Note: there are several ways to solve a system of linear equations.

Example 3:

The height of a rocket, launched from the ground, is given by the expression $h = 45t - 8t^2$, where h is in feet and t is in seconds. Find the height of the rocket at 3 seconds.

Solution:

Replacing t = 3 in the quadratic expression:

$$h = 45t - 8(3)^2$$

$$h = 45(3) - 8(9)$$

$$h = 135 - 72 = 63$$

Then, the height of the rocket is 63 feet at 3 seconds.

Practice Exercises

Rush Road Taxi charges a \$2.45 flat rate in addition to \$0.90 per mile. Mary has no more than \$40 to spend on a ride. Let m be the number of miles Mary can travel.

1. What is the inequality that represents Mary's situation?

   A. $2.45 - 0.90m \leq 40$

   B. $2.45 + 0.90m \geq 40$

   C. $2.45 + 0.90\ m \leq 40$

   D. $2.45 + 0.90m < 40$

2. How many miles can Mary travel without exceeding her budget?

   A. 47.17 miles

   C. 43.80 miles

   B. 41.72 miles

   D. 42.34 miles

3. Suppose that Rush Road Taxi charges a \$4.80 flat rate. How many miles can Mary travel without exceeding her budget?

   A. 39.11 miles

   C. 39.50 miles

   B. 40 miles

   D. 45 miles

Sara buys 5 pretzels and 6 coffees in a shop and the cost is \$36.50. Joe buys 10 pretzels and 8 coffees in the same shop and they cost \$57.
Let x be the cost of each pretzel and y the cost of each coffee.

4. What are the simultaneous linear equations that represent the problem?

   A. 5x + 6y = 57 and 10x + 8y = 36.50

   B. 6x + 5y = 36.50 and 8x + 10y = 57

   C. 6x + 8y = 36.50 and 10x + 5y = 57

   D. 5x + 6y = 36.50 and 10x + 8y = 57

5. What is the cost of each pretzel?

   A. \$3.00

   B. \$3.50

   C. \$2.50

   D. \$2.00

6. What is the cost of each coffee?

   A. \$4.00

   B. \$4.50

   C. \$5.00

   D. \$6.25

7. Solve the following inequality.

   10x + 50 < 20x

   A. x < 5

   B. x > 5

   C. x > - 5

   D. x < 10

8. Let $k = 0.5m^2 + 2$. What is the value of k when m = 6?

   A. 5

   B. 18

   C. 22

   D. 20

$$A - B = 0$$
$$3A - B = 16$$

9. What is the value of A?

A. 8

B. 16

C. 12

D. 4

10. What is the value of A + B?

A. 24

B. 12

C. 8

D. 16

Answer Keys:

1) C
2) B
3) A
4) D
5) C
6) A
7) B
8) D
9) A
10) D

Lesson 2: Understanding and Using Function Notation

A **function** is a relation with each possible input value paired to exactly one output value.

The input values make up the domain, and the output values make up the range. Often, functions are represented as equations. **Function notation** names functions and its variable.

Function Notation

$$f(x) = 2x + 3$$

The notation f(x) defines a function named f. This is read as "y is a function of x." The letter x represents the input value, or **independent variable**. The expression f(x) replaces y and represents the output value, or **dependent variable**.

So, when we write $f(x)=2x+3$, we are saying that the function f takes the input x and produces an output value represented by y, which is determined by the expression 2x+3.

If the function is evaluated at some value, then that number is substituted for the variable in the name. For example, if function f(x) is evaluated at x = 5, then the name is written f(5).

Example 1:

Use function notation to represent a function where the input is the age of a child and the output is its height.

Solution:

The function gives the height of a child, so we can call it *h*. The input is the age of the child, so call it *a*. Thus, the function would be $h(a)$.

Example 2:

A function N(t) gives the number of wolves, N, in a region in years t as a number. What does N(3) = 43 represent?

Solution:

The input of the function is the number of years as a number. The input of N(3) = 43 is 3 and represents 3 years. The output is the number of wolves in the region, which is 43. Thus, the function is saying there were 43 wolves in the region after three years.

Example 3:

Given $f(x) = 4x - 8$, evaluate f(6).

Solution:

Substitute the input value, 6, for x in the function and simplify.

$$f(6) = 4(6) - 8$$

$$f(x) = 24 - 8 = \mathbf{16}$$

Practice Exercises

A dairy spends $9,000 per year to maintain its barns and equipment. It costs $1,500 per year to feed and care for each dairy cow.
Let *x* be the number of dairy cows and *f(x)* be the total yearly expense, in dollars.

1. *Which function gives the total yearly expense as a function of the number of dairy cows?*

   A. $f(x) = 9{,}000x + 1{,}500$
   B. $f(x) = 1{,}500x + 9{,}000$
   C. $f(x) = 9{,}000x - 1{,}500$
   D. $f(x) = 1{,}500x - 9{,}000$

2. *Which function notation expresses the total expense if the dairy has 75 cows?*

   A. $f(x) = 75$
   B. $f(75) = 75$
   C. $f(75)$
   D. $f(x) + 75$

3. *What is the value of* $f(50)$*?*

   A. 10,500
   B. 10,550
   C. 75,000
   D. 84,000

4. *What is the value of* $f(0)$*?*

   A. 1,500
   B. 10,500
   C. 9,000
   D. 7,500

Victoria's coffee stand sells cups of coffee for \$2.50 per cup. She spends \$48 on supplies. She can sell up to 60 cups with the supplies she has purchased.
Let x be the number of cups of coffee and $f(x)$ be the profit in dollars.

5. $Which\ function\ notation\ gives\ Victoria's\ profit?$

   A. $f(x) = 2.50x - 48$

   B. $f(x) = 2.50x + 48$

   C. $f(x) = 48x - 2.50$

   D. $f(x) = 48x + 2.50$

6. $What\ is\ Victoria's\ profit\ if\ she\ sells\ 30\ cups\ of\ coffee?$

   A. \$123

   B. \$27

   C. \$80.50

   D. \$48

7. $What\ is\ her\ maximum\ profit?$

   A. \$198

   B. \$108

   C. \$102

   D. \$150

8. $If\ f(x) = 2x^2 - 5x + 6, what\ is\ f(-1)?$

   A. 7

   B. 3

   C. – 3

   D. 13

9. Paul's class had a math test where the grades were between 0 and 100.Let N(x) represent the number of students whose grade on the exam was x.What does the statement N(87) = 9 mean?

   A. There are 87 students whose grade on the exam was 9.

   B. There are 9 students whose grade on the exam was 87.

   C. The average grade of the class was 87.

   D. The average grade of the class was 9.

10. If $g(x) = 0.5x + 10$, what is $g(0) + g(1)$?

A. 20.5

B. 22

C. 21

D. 11.5

Answer Keys:

1) B
2) C
3) D
4) C
5) A
6) B
7) C
8) D
9) B
10) A

Lesson 3: Rearranging Formulas to Highlight a Quantity Of Interest

A **formula** is a rule in the form of a statement expressed as mathematics symbols to help solve a problem. When we know the value of one quantity, we can find the value of the other using the formula.

Example 1:

Given the following formula $S = \frac{A+B}{L}$, solve for B.

Solution:

Isolate B using the properties of algebra.

$$S = \frac{A+B}{L}$$

$$SL = A + B$$

$$A + B = SL$$

$$\boldsymbol{B = SL - A}$$

Example 2:

Solve for *a*.

$$R = 2a^2 - c$$

Solution:

Isolate *a* using the properties of algebra.

$$R = 2a^2 - c$$

$$2a^2 - c = R$$

$$2a^2 = R + c$$

$$a^2 = \frac{R + c}{2}$$

Take the square root of both sides:

$$\sqrt{a^2} = \sqrt{\frac{R + c}{2}}$$

$$a = \sqrt{\frac{R + c}{2}}$$

Practice Exercises

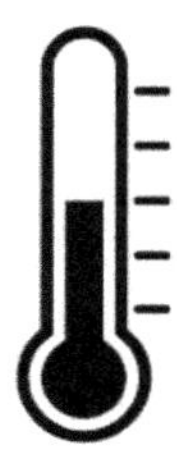

The temperature conversion formula from Celsius to Fahrenheit is:

$$\mathbf{°F = \frac{9}{5}°C + 32}$$

Where C is the temperature in degrees Celsius (°C) and F is the temperature in degrees Fahrenheit (°F)

1. What is the temperature conversion formula from Fahrenheit to Celsius?

   A. $°C = \frac{5}{9}°F - 32$

   B. $°C = \frac{5}{9}°F + 32$

   C. $°C = \frac{5}{9}(°F - 32)$

   D. $C = \frac{9}{5}(°F - 32)$

2. The human body temperature is generally 37°C. What is the body temperature in Fahrenheit degrees?

   A. 98.6 °F

   B. 34.6 °F

   C. 52.6 °F

   D. 102 °F

3. The boiling point of water is 212 °F. What is the boiling point of water in Celsius degrees?

A. 324 °C

B. 200 °C

C. 100 °C

D. 414 °C

4. What is the temperature in Fahrenheit of 0 °C?

A. – 32 °F

B. 32 °F

C. 64 °F

D. 0°F

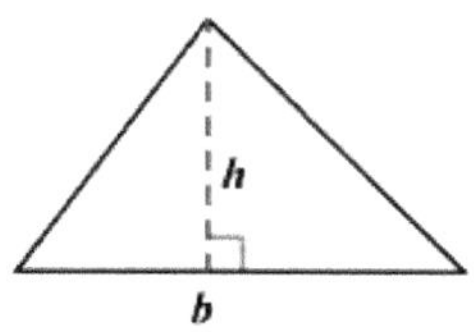

The formula for the area of a triangle is equal to half the product of its base and height.

$$A = \frac{b \cdot h}{2}$$

A is the area of the triangle, b is the base and h is its height.

5. What is the formula for the height of a triangle?

A. $h = \frac{2A}{b}$

B. $h = 2Ab$

C. $h = \frac{2b}{A}$

D. $h = 2A - b$

6. What is the formula for the base of a triangle?

A. $b = \frac{2h}{A}$

B. $b = 2Ah$

C. $b = \frac{A}{2h}$

D. $b = \frac{2A}{h}$

7. What is the area of a triangle whose height and base are 15 ft. and 9 ft. respectively?

A. 135 ft^2

B. 67.5 ft^2

C. 62.5 ft^2

D. 140 ft^2

8. What is the base of a triangle whose height and area are 12 ft. and 144 ft^2, respectively?

A. 12 ft.

B. 36 ft.

C. 24 ft.

D. 18 ft.

9. For the formula P = 3k – 4m, what is the value of P when k = 3.5 and m = 1.5?

A. 4.5

B. 4

C. 3.5

D. 16.5

10. For the formula $H = 3k^2$, what is the value of k when H = 27?

A. 9

B. 6

C. 4.5

D. 3

Answer Keys:

1) C
2) A
3) C
4) B
5) A
6) D
7) B
8) C
9) A
10) D

REFLECTION ON LEARNING

Answer the following reflection questions and feel free to discuss your responses with your teacher or a classmate.

1- What math ideas and principles did you learn in this chapter?

2- What new math concepts did you learn?

3- What procedures or methods did you practice in this chapter?

4- What aspect of this chapter is still not 100% clear to you?

5- What else do you want your teacher to know?

CHAPTER 4
GEOMETRY AND MEASUREMENT

Lesson 1: Solving Real-World Problems Involving Volume and Surface Area

Volume is the space occupied within the boundaries of an object in three-dimensional space.

Volume of Three – Dimensional Shapes

Name of the solid	Figure	Volume	Nomenclature
Cube	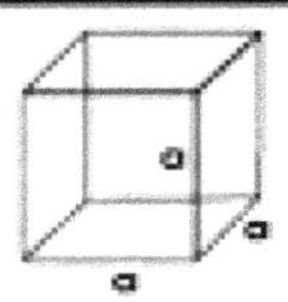	a^3	a : side of the cube
Cuboid	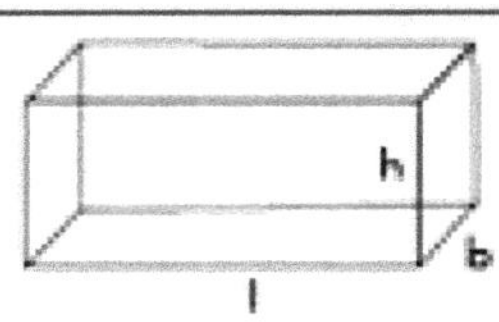	lbh	l : length b : width h : height
Cone	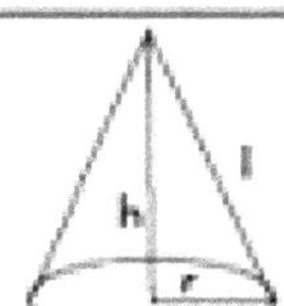	$\frac{1}{3}\pi r^2 h$	r : radius of the base h : height l : slant height
Cylinder	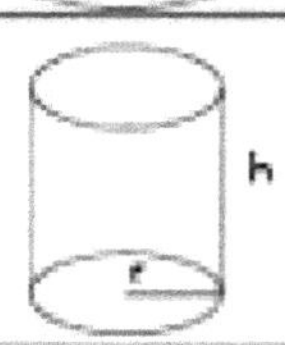	$\pi r^2 h$	r : radius of the base h : height
Sphere	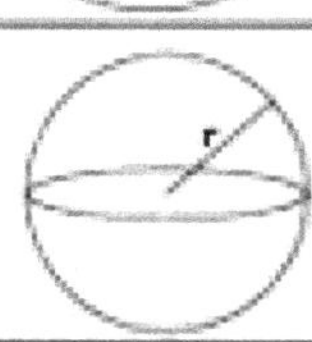	$\frac{4}{3}\pi r^3$	r : radius

The **surface area** of a three-dimensional shape is the total area of all of its faces. To find the surface area of a shape, we find the area of each face and add them together.

Name of the Solid	Figure	Lateral/Curved Surface Area	Total Surface Area	Nomenclature
Cube	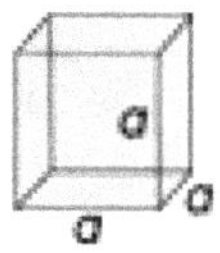	$4a^2$	$6a^2$	a : side of cube
Cuboid	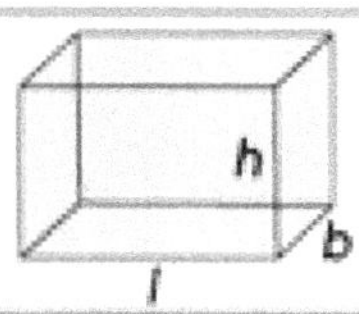	$2h\ (l + b)$	$2(lb + bh + hl)$	l : length b : breadth h : height
Cone	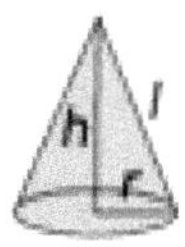	πrl	$\pi r(l + r)$	r : radius of base h : height l : slant height
Cylinder	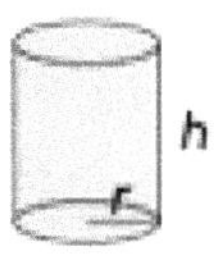	$2\pi rh$	$2\pi r(r + h)$	r : radius of base h : height
Sphere		$4\pi r^2$	$4\pi r^2$	r : radius

Example 1:

Helen has a cylindrical water bottle with a base radius of 2 inches and a height of 5 inches. What is the volume of water that the bottle can store? (Use $\pi = 3.14$)

Solution:

The volume of a cylinder is

$$V = \pi r^2 h$$

Replacing the values of r and h into the formula, we get:

$$V = 3.14 \cdot (2\ in)^2 (5\ in)$$

$$V = 62.8\ in^3$$

Example 2:

Nick built a wooden, cubic toy box for his daughter. Each side of the box measures 3 feet. How many square feet of wood did he use to build the box?

Solution:

The surface area of a cube is $S = 6a^2$ where *a* is the side length of the cube. Replacing a = 3 ft. into the formula, we get:

$$S = 6(3\,ft)^2$$

$$S = 6 \cdot 9\,ft^2 = 54ft^2$$

Thus, Nick used 54 square feet of wood.

Practice Exercises

The radius of a soccer ball is 4 inches.
Use $\pi = 3.14$

1. What is the volume of the soccer ball?

   A. 50.24 in^3
   B. 267.95 in^3
   C. 200.96 in^3
   D. 100.48 in^3

2. What is the surface area of the soccer ball?

   A. 200.96 in^2
   B. 50.25 in^2
   C. 37.68 in^2
   D. 75.68 in^2

3. If the surface area of the soccer ball is 153.86 square inches, what is the radius of the ball?

   A. 4 inches
   B. 4.5 inches
   C. 5 inches
   D. 3.5 inches

An **open** box in the shape of a cube has 6.5 inches side length.

4. What is the volume of the box?

   A. 42.25 in^3
   B. 19.50 in^3
   C. 78.00 in^3
   D. 274.63 in^3

5. What is the surface area of the box?

   A. 253.10 in^2
   B. 211.25 in^2
   C. 253.5 in^2
   D. 250.5 in^2

6. If the side length of the box is 10 inches, what is the surface area?

   A. 60 in^2
   B. 600 in^2
   C. 500 in^2
   D. 400 in^2

7. A swimming pool is 27 feet long, 20 feet wide, and 7 feet deep. What is the volume of the pool?

   A. 1,730 ft^3
   B. 3,780 ft^3
   C. 2,916 ft^3
   D. 3,563 ft^3

8. A paint roller is cylindrical in shape. It has a diameter of 1.5 inches and a width of 13 inches. What is the area painted by the roller when it makes one revolution? (Use $\pi = 3.14$)

   A. 91.85 in^2
   B. 183.70 in^2
   C. 122.46 in^2
   D. 61.23 in^2

The solid below is made up of cubes, each of which has 5 inches side length.

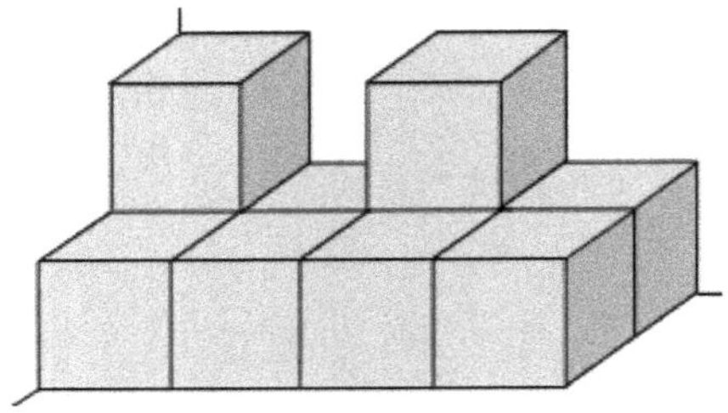

9. What is the volume of one cube?

A. 75 in^3

B. 150 in^3

C. 125 in^3

D. 105 in^3

10. What is the volume of the solid figure?

A. 1,000 in^3

B. 1,200 in^3

C. 1,100 in^3

D. 1,250 in^3

Answer Keys:

1) B
2) A
3) D
4) D
5) C
6) B
7) B
8) D
9) C
10) D

Lesson 2: Applying the Pythagorean Theorem In Real-World Contexts and on the Coordinate Plane

The Pythagorean Theorem is a relationship between the sides of a right triangle. In a right triangle, the sides are called the legs and the hypotenuse.

The two legs meet at a 90° angle and the hypotenuse is the side opposite the right angle.

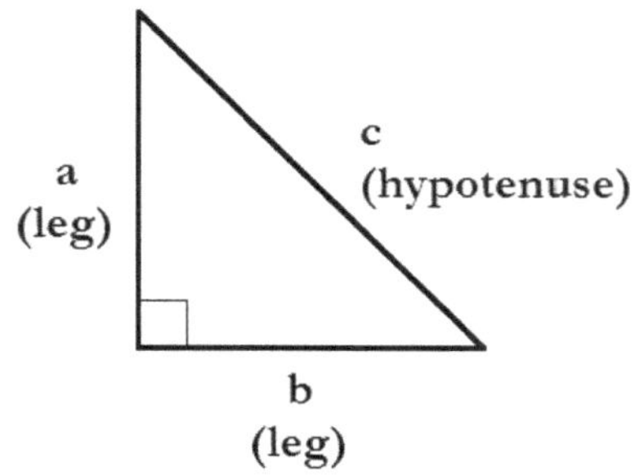

The Pythagorean Theorem states the following:

In a right triangle, the square of the hypotenuse is equal to the sum of the squares of the two legs.

This theorem can be written in one equation.

The Pythagorean Theorem

$$a^2 + b^2 = c^2$$

Example 1:

James wants to replace the net on his basketball hoop. The hoop is 10 feet high. James places his ladder 5 feet from the base of the hoop. How long must the ladder be to reach the hoop?

Solution:

Draw a picture to represent the problem:

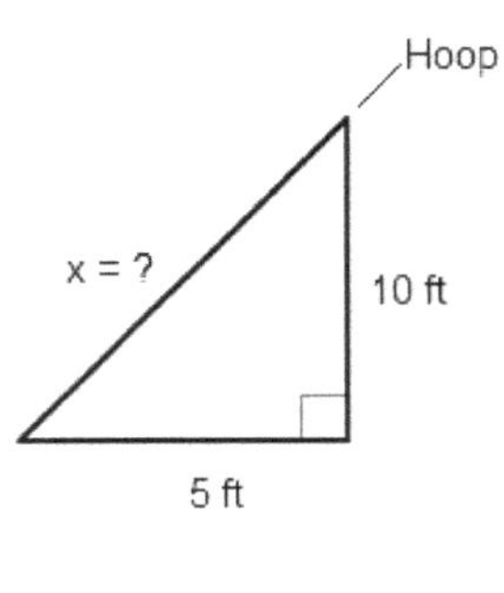

Let x be the length of the ladder. Applying the Pythagorean Theorem, we can find the value of x.

$$x^2 = (10ft)^2 + (5ft)^2$$

$$x^2 = 100ft^2 + 25ft^2$$

$$x^2 = 125ft^2$$

Take the square root of both sides:

$$\sqrt{x^2} = \sqrt{125ft^2}$$

$$x = 11.18\ ft$$

Then, the length of the ladder must be **11.18 ft.** to reach the hoop.

We can apply the Pythagorean Theorem on the coordinate plane to solve real-world problems.

Example 2:

The coordinates of the vertices of a triangle are A = (3, 2), B = (6, 2), and C = (3, 6). Find the length of side BC.

Solution:

Plot the vertices on the coordinate plane and draw the triangle ABC.

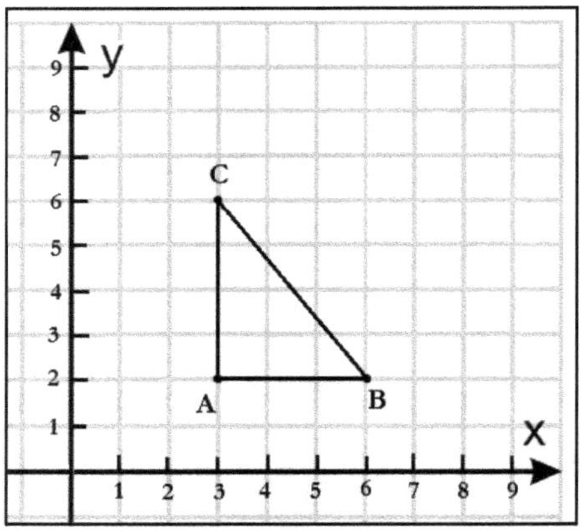

Based on the drawing, we can determine that the triangle ABC is a right triangle. Notice that $\overline{AB} = 3$ and $\overline{AC} = 4$. The side BC is the hypotenuse of the triangle ABC. Apply the Pythagorean Theorem to find the length of side BC.

$$(BC)^2 = (AB)^2 + (AC)^2$$

$$(BC)^2 = (3)^2 + (4)^2$$

$$(BC)^2 = 9 + 16$$

$$(BC)^2 = 25$$

Take the square root of both sides:

$$(BC)^2 = 25$$

$$\sqrt{(BC)^2} = \sqrt{25} \quad \Rightarrow \quad BC = 5$$

Practice Exercises

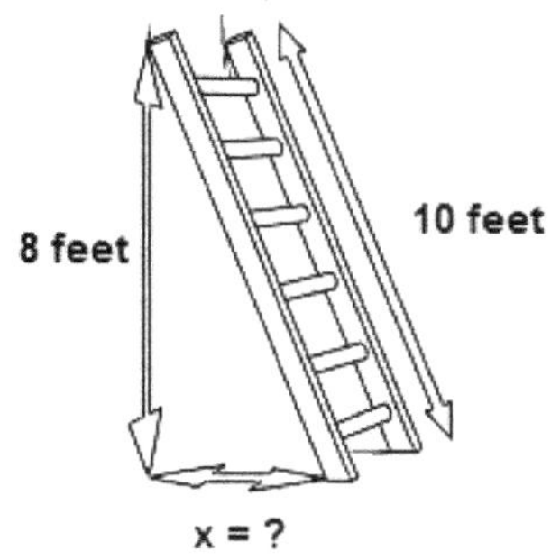

A ladder that is 10 feet long is placed on the ground in such a way that it touches the top of a vertical wall 8 feet high. Let x be the distance of the foot of the ladder from the bottom of the wall.

1. Which expression gives the distance of the foot of the ladder from the bottom of the wall?

   A. $x^2 + 10^2 = 8^2$

   B. $x = \sqrt{10^2 + 8^2}$

   C. $x = \sqrt{10^2 - 8^2}$

   D. $x^2 + 8 = 10$

2. What is the distance of the foot of the ladder from the bottom of the wall?

   A. 6 feet

   B. 7 feet

   C. 13 feet

   D. 18 feet

3. If the length of the ladder is 15 feet, what is the distance of the foot of the ladder from the bottom of the wall? (Round your answer to the nearest hundredth.)

   A. 17 feet

   B. 23 feet

   C. 15.45 feet

   D. 12.69 feet

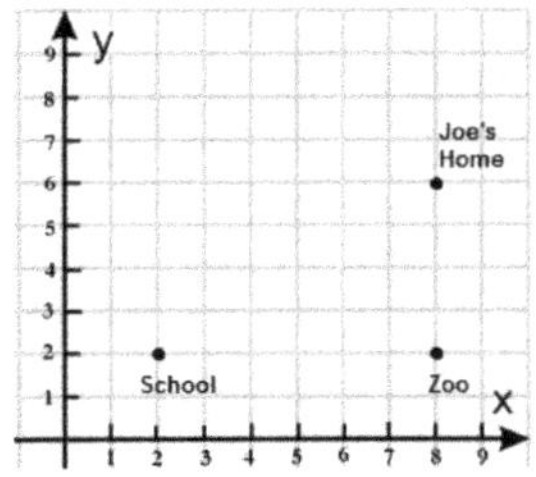

The coordinate plane shows some places near Joe's home.

4. What is the distance between the school and the zoo?

A. 8

B. 2

C. 6

D. 7

5. What is the distance between the zoo and Joe's home?

A. 6

B. 2

C. 5

D. 4

6. How far is the school from Joe's home? (Round your answer to the nearest hundredth)

A. 8.56

B. 7.45

C. 6.62

D. 7.21

7. Suppose the coordinates of Joe's home are (8, 8). How far is the school from Joe's home?

A. 10

B. 5.29

C. 8.25

D. 9

8. Two legs of a right rectangle measure 1 foot and 5 feet. What is the length of the hypotenuse?

A. 6 ft.

B. $\sqrt{24}$ ft.

C. 2 ft.

D. $\sqrt{26}$ ft.

9. In a right triangle, the length of the base is 5 inches, and the hypotenuse is 13 inches. What is the length of the missing side?

A. $\sqrt{124}$ in.

B. 17 in.

C. $\sqrt{324}$ in.

D. 12 in.

10 Which expression can we use to find the value of y?

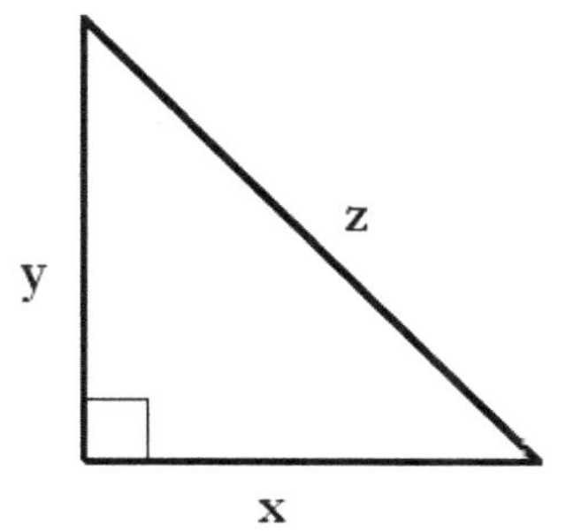

A. $y = \sqrt{x^2 + z^2}$

B. $y = \sqrt{x^2 - z^2}$

C. $y = \sqrt{x^2 + y^2}$

D. $y = \sqrt{z^2 - x^2}$

Answer Keys:

1) C	4) C	7) A	10) D
2) A	5) D	8) D	
3) D	6) D	9) D	

Lesson 3: Solving Problems Involving Similarity and Congruence

Similar figures are the same shape as the real objects they represent but not the same size. The **scale factor** is **the ratio** of corresponding side lengths of similar figures. It's a measure of how much the dimensions of one figure have been scaled up or down to create the other figure.

Similar Figures

If one shape or geometric figure can become another using rotations, translations, or reflections, then the figures are **congruent.** In other words, figures are congruent **when they have the same shape and size**, but we are allowed to flip, slide or turn them.

Congruent Figures

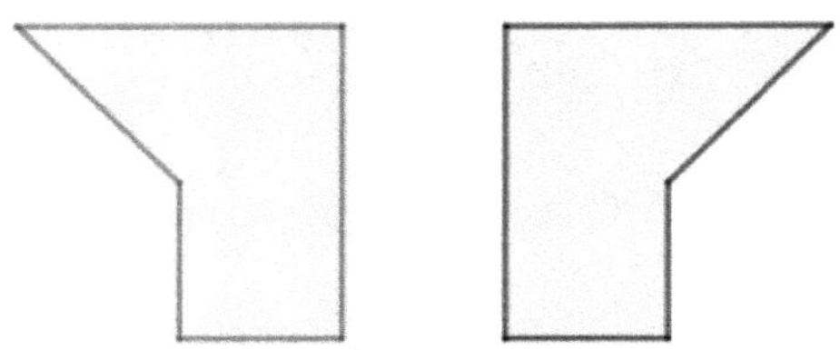

Example 1:

The following rectangles are similar. What is x?

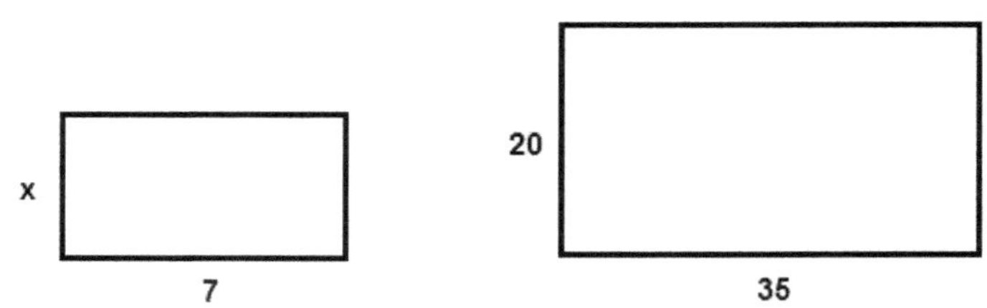

Solution:

Since the rectangles are similar, we set up the proportion between corresponding side lengths.

$$\frac{x}{7} = \frac{20}{35}$$

Solve for x:

$$\frac{x}{7} = \frac{20}{35}$$

$$x = \frac{20(7)}{35}$$

$$x = \frac{140}{35} = 4$$

Then, **x = 4**

Example 2:

Which shape is not congruent to the other two?

Solution:

Notice that Figures 1 and 2 have the same size, area, angles, and side lengths. Also, Figure 2 is a rotation of Figure 1. Figure 3 is similar to the other figures. Therefore, Figure 3 is not congruent to the other figures.

Practice Exercises

The following triangles are similar:

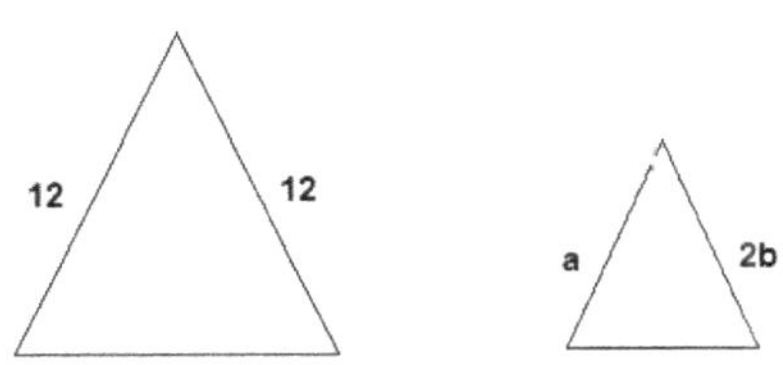

1. What is the value of a?

   A. 5.5

   B. 6

   C. 4.8

   D. 5

2. What is the value of b?

A. 2.4

B. 4

C. 4.6

D. 3

3. What is the value of a + 2b?

A. 10

B. 8

C. 12

D. 9.6

The following figures are congruent.

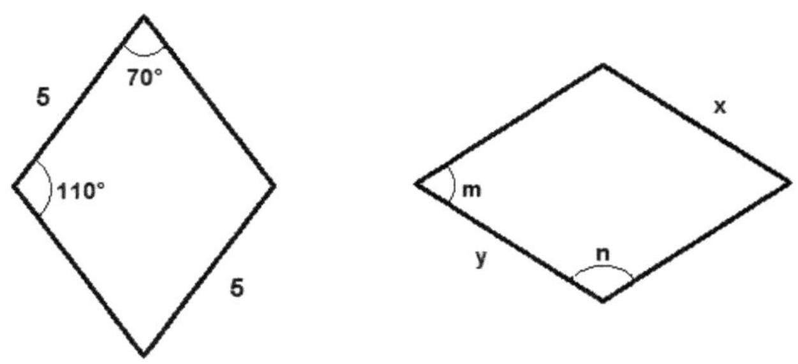

4. What is x?

A. 5

B. 70°

C. 10

D. 110°

5. What is the measure of the angle m?

A. 110°

B. 70°

C. 180°

D. 40°

6. What is the value of x – y?

A. 10

B. 5

C. 15

D. 0

7. What is the value of m + n?

A. 110°

B. 40°

C. 180°

D. 70°

8. Which triangles are congruent?

Look at the following triangles.

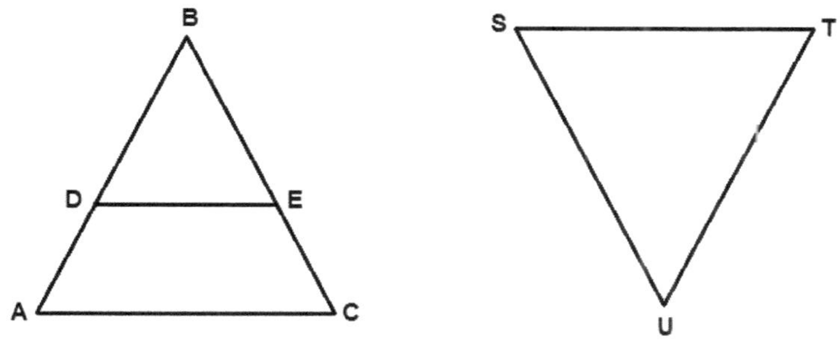

A. Triangle ABC and Triangle DBE

B. Triangle DBE and Triangle STU

C. Triangle STU and Triangle ABC

D. None of the above

9. Which triangles are similar?

A. Triangle STU and Triangle DBE

B. Triangle ABC and Triangle DBE

C. Triangle STU and Triangle ABC

D. None of the above

10. If the area of the triangle ABC is 12 in^2, what is the area of the triangle STU?

A. 6 in^2

B. 12 in^2

C. 10 in^2

D. 8 in^2

Answer Keys:

1) C
2) A
3) D
4) A
5) B
6) D
7) C
8) C
9) B
10) B

Lesson 4: Understanding the Concept of Density Based on Area and Volume

Density is a measure of the amount of a substance contained in a certain volume. In various contexts, we will find the use of density to represent a range of quantities including population, cost, and surface area. We can apply the concept of density to both area and volume when modeling situations and events.

Example 1:

New York City has 8.3 million people and an area of 302 square miles. Paris has 2.1 million people and an area of 40 square miles. Which city has a greater population density?

Solution:

Find the population density of New York City.

$$Density = \frac{people}{area}$$

$$Density = \frac{8{,}300{,}000}{302\ sq\ mil.} = 27{,}483\ people\ per\ square\ mile$$

Find the population density of Paris.

$$Density = \frac{2{,}100{,}000}{40\ sq\ mil.} = 52{,}500\ people\ per\ square\ mile$$

Compare both population densities. Then, Paris has a greater population density than New York City.

Practice Exercises

The following table shows the land areas and populations of four states.

State	Populations (millions)	Area (square miles)
Arkansas	3.02	52,035
Kentucky	4.45	39,486
Mississippi	3.10	46,923
Tennessee	6.70	41,235

1. What is the population density of Kentucky? (Round your answer to the nearest tenth.)

   A. 88.7 people per square mile.

   B. 112.7 people per square mile.

   C. 66.1 people per square mile.

   D. 100.5 people per square mile.

2. What is the population density of Tennessee? (Round your answer to the nearest tenth.)

   A. 57.9 people per square mile

   B. 113.1 people per square mile

   C. 162.5 people per square mile

   D. 150.7 people per square mile

3. Which state has the greatest population density?

   A. Arkansas

   B. Mississippi

   C. Kentucky

   D. Tennessee

4. Which state has the lowest population density?

   A. Mississippi

   B. Tennessee

   C. Arkansas

   D. Kentucky

Shanghai has 26.3 million people and an area of 2,488 square miles. New York has 8.3 million people and an area of 302 square miles.

5. What is the population density of Shanghai? (Round your answer to the nearest tenth.)

   A. 1,057.1 people per square mile

   B. 10,570.7 people per square mile

   C. 105,707 people per square mile

   D. 13,665.6 people per square mile

6. Which city has the greater population density?

   A. New York

   B. Shanghai

   C. Cannot be determined

7. How many times as great is the population density of New York as Shanghai?

   A. About 5 times as great

   B. About 3 times as great

   C. About 2 times as great

   D. About the same

8. The population density of Snowville is 73 people per square mile. 3,650 people are living in Snowville. How many square miles is Snowville?

   A. 45 square miles

   B. 60 square miles

   C. 55 square miles

   D. 50 square miles

In Chemistry, density (d) is a physical property found by dividing the mass of an object (m) by its volume (v).

9. Which would be the formula for the density?

   A. $d = \frac{v}{m}$

   B. $d = mv$

   C. $d = \frac{m}{v}$

   D. $d = \frac{1}{mv}$

10. What is the density of an object that has a volume of 100 cubic centimeters and a mass of 50 grams?

    A. 2 g/cm^3

    B. 0.5 g/cm^3

    C. 5 g/cm^3

    D. 50 g/cm^3

Answer Keys:

1) B
2) C
3) D
4) C
5) B
6) A
7) C
8) D
9) C
10) B

REFLECTION ON LEARNING

Answer the following reflection questions and feel free to discuss your responses with your teacher or a classmate.

1- What math ideas and principles did you learn in this chapter?

2- What new math concepts did you learn?

3- What procedures or methods did you practice in this chapter?

4- What aspect of this chapter is still not 100% clear to you?

5- What else do you want your teacher to know?

CHAPTER 5
DATA ANALYSIS, STATISTICS AND PROBABILITY

Lesson 1: Understanding and Applying the Concept of Probability

Probability is the chance that something will happen. It's calculated by dividing the number of favorable outcomes by the total number of outcomes. This means that probability is **always a number between 0 and 1.**

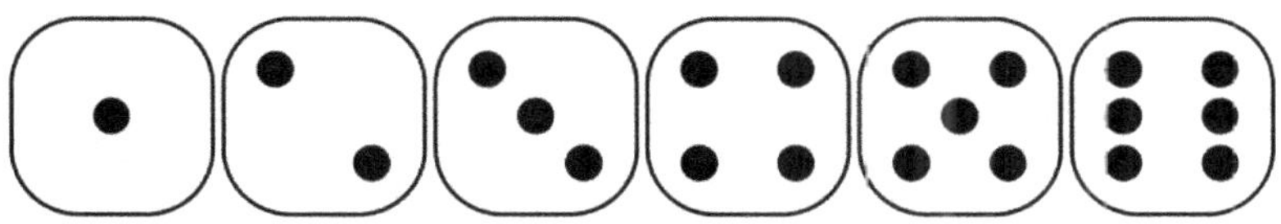

An event with a probability of 0 means it can never happen, as there are no favorable outcomes. An event with a probability of 1 will certainly happen because there are the same number of favorable outcomes as there are total numbers of outcomes.

Probabilities can be expressed in terms of ratios. Since any ratio can be turned into a fraction, decimal, or percent, we can also turn any probability into a fraction, decimal, or percent. For example, the probability that a fair coin lands on tails is ½. This probability can be expressed in several forms:

$$P = \frac{1}{2}, \qquad P = 0.5, \qquad P = 50\%$$

Example 1:

Find the probability of rolling an odd number on a regular 6-sided die, and the probability of rolling 7 on a regular 6-sided die.

Solution:

For the probability of rolling an odd number, notice there are 3 odd numbers on a die (1, 3, 5), and there are a total of 6 possible outcomes (1, 2, 3, 4, 5, 6).

Write the expression for the probability of rolling an odd number.

$$Probability = \frac{Number\ of\ ways\ it\ can\ happen}{Total\ number\ of\ outcomes} = \frac{3}{6} = \mathbf{\frac{1}{2}}$$

For the probability of rolling 7, note that the number 7 does not appear on a regular 6-sided die. It is an impossible event. Then, the probability of rolling a 7 on a regular 6-sided die is **0.**

Practice Exercises

1. Which of the following events has a probability that equals 1?

   A. A six will come up if a die is tossed.

   B. A head will come up if a coin is tossed.

   C. I will eat bacon next week.

   D. The sun will rise tomorrow.

2. What is the probability of getting a number less than 4 when a die is rolled?

   A. 40%

   B. 67%

   C. 50%

   D. 10%

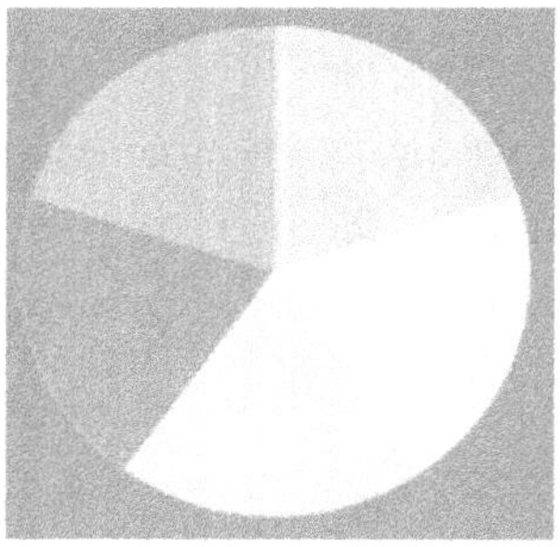

A spinner wheel has 5 equal sectors colored blue, green, red, brown, and yellow.

3. What is the probability of landing on red after spinning the spinner?

   A. 20%

   B. 5%

   C. 10%

   D. 25%

4. What is the probability of landing on white after spinning the spinner?

   A. 10%

   B. 20%

   C. 1%

   D. 0%

5. If the spinner wheel has six equal sectors colored blue, green, red, brown, yellow, and orange, what is the probability of landing on blue after spinning the spinner?

A. 15%
B. 16.7%
C. 18.5%
D. 12.4%

6. Which of these numbers cannot be a probability?

A. 120%
B. 0.01
C. 8/9
D. 0.99

7. A number from 1 to 20 is chosen at random. What is the probability of choosing an even number?

A. 0.10
B. 0.20
C. 0.25
D. 0.50

8. The probability of an event is always a number that is:

A. Between 0% and 50%
B. Between 50% and 100%
C. Between 0% and 100%
D. Between 0% and 1%

9. The probability that an event never occurs is:

A. 1
B. 0.50
C. 0.25
D. 0

10. Bryan rolls a fair dice, with sides labelled M, P, U, S, O and T. What is the probability that the dice lands on a vowel?

A. 1/3
B. 1/6
C. 2/3
D. 1/2

Answer Keys:

1) D
2) C
3) A
4) D
5) B
6) A
7) D
8) C
9) D
10) A

Lesson 2: Using 2-Way Tables to Interpret Bivariate Data

A two-way table is a visual representation of the possible relationships between two sets of categorical data. The categories are labeled at the top and the left side of the table, with **the frequency** (count) information appearing in the interior cells of the table.

Example of a Two-Way Table

Gender	Positive opinion	Negative Opinion	Total
Male	30	20	50
Female	20	18	38
Total	50	38	88

Example 1:

The following two-way frequency table shows the preference of cats or dogs in a classroom:

Preference	Male	Female
Cats	15	24
Dogs	32	11
No preference	5	8

How many students prefer dogs?

Solution:

Add up the number of students in the Dogs row:

Preference	Male	Female	Total
Dogs	32	11	43

Then, the number of students that prefer dogs is 32 + 11 = **43**

Practice Exercises

The following two-way table shows a data set about what students eat for lunch. One categorical variable is each child's gender; the other is the type of lunch they have.

	Male	Female
Cooked Food	49	37
Packed Lunch	65	58

1. How many students prefer packed lunch?

   A. 65

   B. 97

   C. 123

   D. 103

2. How many students prefer cooked food?

   A. 86

   B. 114

   C. 49

   D. 95

3. How many students were asked?

   A. 114

   B. 187

   C. 209

   D. 231

The following two-way table shows some information about the number of bookcases a factory makes in one month.

	Small	Medium	Large	Total
Pine	18	20	15	
Oak	23	B	18	73
Yew	A	13	19	C
Total	82			

4. What is A?

A. 82

B. 39

C. 41

D. 43

5. What is B?

A. 32

B. 35

C. 27

D. 61

6. What is C?

A. 86

B. 82

C. 71

D. 73

7. How many bookcases were made in a month?

A. 251

B. 177

C. 210

D. 199

The following two-way table shows the preference of sports in a college.

	Male	Female	Total
Swimming	A	91	268
Basketball	168	B	300
Football	C	50	
Tennis	143	156	
Total	618		

8. What is A?

A. 177

B. 164

C. 181

D. 178

9. What is C?

A. 123

B. 176

C. 130

D. 132

10. How many students were asked?

A. 1,047

B. 1,060

C. 967

D. 1,119

Answer Keys:

1) C	4) C	7) D	10) A
2) A	5) A	8) A	
3) C	6) D	9) C	

Lesson 3: Interpreting and Comparing Data Sets, Including Comparisons of Statistical Variability

Center and Spread

When we talk about center and spread, we are referring to the distribution of data or how data is spread across a graph or chart.

The center of a distribution is a measure of what a typical value would be. Two common measures for the center of a distribution are the mean and the median. **The mean** is the average of the numbers. To calculate the mean, we must add all the values and then divide by the total number of values of the data set. **The median**, on the other hand, is the middle of a sorted list of numbers.

The spread of data refers to the variability of the data. The simplest way to find the spread in a data set is to identify **the range**, which is the difference between the highest and lowest values in the set. Additionally, the number that appears the most often is called the **mode**.

An outlier is a portion of data that is an abnormal distance from other points. In other words, it is data that lies outside the other values in the set.

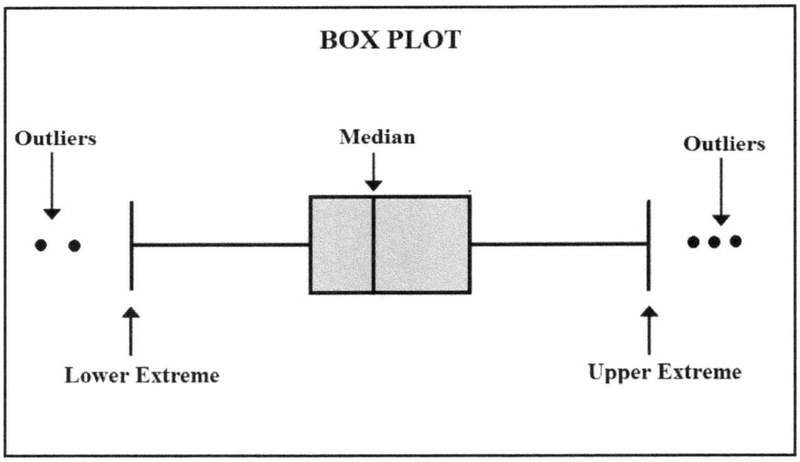

Example 1:

Find the mean, median, mode and range of the following set of data:

12, 13, 15, 9, 6, 21, 17, 19

Solution:

To calculate the mean, add together all of the numbers in the set and then divide the sum by how many numbers there are. (There are 8 numbers.)

$$Mean = \frac{12 + 13 + 15 + 9 + 6 + 21 + 17 + 19}{8} = \frac{112}{8} = 14$$

Then, the mean is **14.** To calculate the median, all of the values are first placed in numerical order.

6, 9, 12, 13, 15, 17, 19, 21

If we have an odd number of values, the median is the middle number. However, if we have an even number of values, the median will be the mean of the two values in the center of the data set. We have an even number of values, so we must calculate the mean of the two central values:

6, 9, 12, **13, 15**, 17, 19, 21

$$Mean = \frac{13 + 15}{2} = \frac{28}{2} = 14$$

Then, the median is **14**. The mode is the value that occurs most often. Notice that no number appears more than once. Thus, there is no mode.

To calculate the range of the data set, subtract the highest value from the lowest value.

Range = 21 – 6 = **15**

Practice Exercises

Malcom normally makes good grades in math class. His test scores are 89, 91, 86, 90, 92, 95, 93 and 68.

1. What is the mean of the scores?

   A. 90
   B. 91
   C. 89
   D. 88

2. What is the median of the scores?

   A. 90.5
   B. 90
   C. 91
   D. 91.5

3. What is the range of the scores?

   A. 9
   B. 12
   C. 6
   D. 27

4. Can the score, 68, be considered an outlier?

A. Yes

B. No

C. There is not enough information given.

Look at the following box plot.

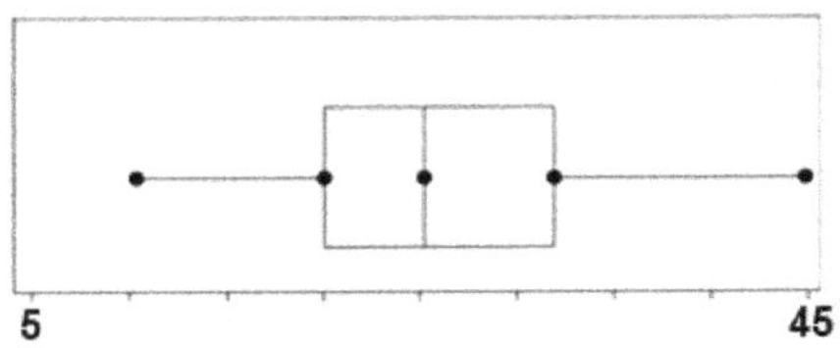

5. What is the median of the data set?

A. 45

B. 25

C. 30

D. 35

6. What is the minimum value of the data set?

A. 5

B. 1

C. 10

D. 15

7. What is the range of the data set?

A. 40

B. 35

C. 30

D. 45

The following data are the number of points scored by a basketball team over its last 10 games.
68, 79, X, 83, 87, 99, 72, 91, 100, and 85.
The mean of the scores is 86.6.

8. What is X?

A. 94

B. 102

C. 110

D. 88

9. What is the median of the scores?

A. 85
B. 87
C. 86
D. 85

10. What is the range of the scores?

A. 34
B. 32
C. 35
D. 29

Answer Keys:

1) D
2) A
3) D
4) A
5) B
6) C
7) B
8) B
9) C
10) A

REFLECTION ON LEARNING

Answer the following reflection questions and feel free to discuss your responses with your teacher or a classmate.

1- What math ideas and principles did you learn in this chapter?

2- What new math concepts did you learn?

3- What procedures or methods did you practice in this chapter?

4- What aspect of this chapter is still not 100% clear to you?

5- What else do you want your teacher to know?

CHAPTER 6
PURE MATHEMATICS

Activity 1

The following table shows the time it takes to ride the elevator to five different floors of a building.

Floor number	1	3	9	27	**B**
Time needed (in seconds)	2.5	**A**	22.5	67.5	**C**

Find A, B, and C.

Activity 2

Complete the following sale receipt.

New Store X1
Some address
мпав, fd, 23rf
Phone +333333333333
SALE

Order #:	010900269558
Sold To:	Walk-In
Order Date:	03/31/17
Order Time:	1:03:23 PM
Sales Person:	Admin Admin
Register:	1
Order type:	Quick Sale

Red Banana is in business	x 3	$66.96
Apple	x 5.55	$33.24
Discount Sale		($5.95)

Subtotal	?
Discount	($5.95)
12 %	?
Total	?

Cash Tendered
CHANGE DUE **$0.00**
Special package

30% off your next order –
if you order through us with code LOCAL!

Activity 3

In an election, Vivian received 248 more votes than Candace. The total number of votes cast was1,544. Find the number of votes each candidate received.

Activity 4

Given $f(x) = 5x^2 - 3x - 2$, evaluate $\frac{f(1)+f(2)}{f(0)}$

Activity 5

Find the probability of choosing a vowel from the alphabet. Write your answer as a fraction.

Practice Problems

1. If 2.5 pounds of meat will serve 6 people, how many pounds will be needed to serve 24 people?

   A. 12 pounds

   B. 10 pounds

   C. 11 pounds

   D. 15.5 pounds

2. Which of the following is true?

   A. $\sqrt{625} = 125$

   B. 0.0909 is an irrational number.

   C. π^2 is an irrational number.

   D. $\sqrt{1} = 0.5$

If 1 Bitcoin = $69,727.90, what is the value of $1,000,000 in Bitcoin? (Round your answer to 2 decimals.)

A. 0.07 Bitcoin

B. 14.34 Bitcoin

C. 7.15 Bitcoin

D. 0.14 Bitcoin

In a department store, a $110 dress is marked, "Save 20%." What is the sale price of the dress?

A. $90.00

B. $88.00

C. $85.75

D. $82.50

5. A taxi charges a flat rate of $2.55 plus an additional $0.95 per mile. If Sarah has at most $50 to spend on the cab ride, how far could she travel?

   A. No more than 19.2 miles

   B. At least 19.2 miles

   C. At least 49.9 miles

   D. No more than 49.9 miles

6. The function W(t) models Frank's weight (in pounds) when he was t years old. What does the statement W(20) = 155 mean?

   A. Frank's weight was 20 pounds.

   B. Frank's weight was 155 pounds 20 years ago.

   C. When Frank was 20 years old, his weight was equal to 155 pounds.

   D. Frank's weight was 155 pounds.

7. What is the surface area of a cube whose volume is 1,000 cubic inches?

   A. 60 in^2

   B. 100 in^2

   C. 600 in^2

   D. 400 in^2

8. The following figures are similar. What is x?

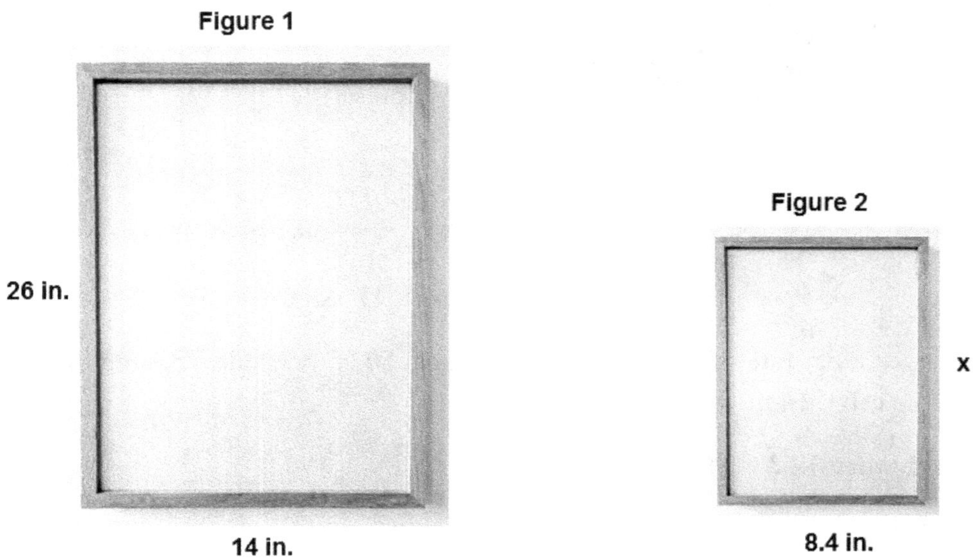

A. 15.6 in.

B. 18.2 in.

C. 13.5 in.

D. 14.2 in.

9. The following triangle is a right triangle. What is m?

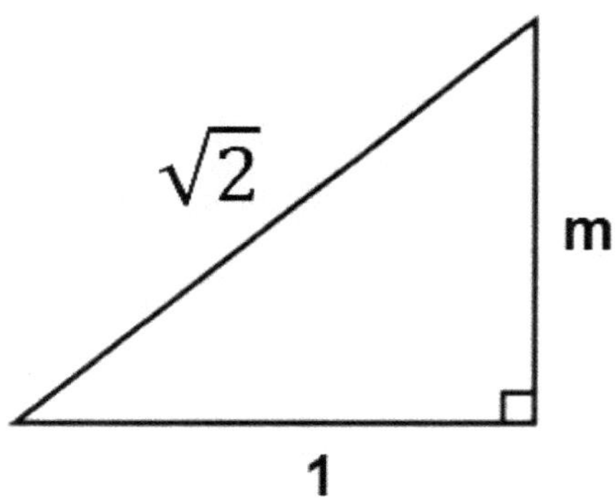

A. 2

B. 1.5

C. 1

D. $\sqrt{3}$

10. The scores obtained by a student in six math tests are given below:

Test	1	2	3	4	5	6
Score obtained (points)	71	68	73	80	66	75

What is the probability of the student getting more than 70 points on a math test?

A. 60%

B. 70%

C. 66.7%

D. 75%

Answer Keys:

Activity 1

A = 7.5, B = 81, C = 202.5

Activity 2

Subtotal = $100.20

12% tax = $11.31

Total = $105.56

Activity 3

Vivian = 896 votes, Candace = 648

Activity 4

$$\frac{f(1) + f(2)}{f(0)} = -6$$

Activity 5

Probability = 5/26

Practice problems

1) B
2) C
3) B
4) B
5) D
6) C
7) C
8) A
9) C
10) C

REFLECTION ON LEARNING

Answer the following reflection questions and feel free to discuss your responses with your teacher or a classmate.

1- What math ideas and principles did you learn in this chapter?

2- What new math concepts did you learn?

3- What procedures or methods did you practice in this chapter?

4- What aspect of this chapter is still not 100% clear to you?

5- What else do you want your teacher to know?

PRACTICE TEST 1

Read each question and choose the best answer. You have 90 minutes to answer 40 questions.

1. What is x in the following proportion?

$$\frac{531\ cars}{x} = \frac{59\ cars}{3\ hours}$$

A. 18 hours

B. 27 cars

C. 21 cars

D. 27 hours

2. Sharon got 3 problems wrong for every 7 problems right on a math test. How many problems did Sharon get wrong if there were 50 problems altogether on the math test?

A. 21

B. 15

C. 20

D. 22

3. For every \$24 that Richard makes, he spends \$16. The rest goes into his savings account. If his weekly take-home pay is \$600, how much does he save each week?

A. \$400

B. \$250

C. \$200

D. \$150

4. Which of the following is an irrational number?

A. $\sqrt{1} + \sqrt{16}$

B. $\sqrt{121}$

C. 0.511…

D. 3/4

5. Which of the following is **not** an irrational number?

A. π

B. $\sqrt{19}$

C. $\sqrt[3]{2}$

D. $\sqrt{100}$

6. Every minute, 2.5 gallons of water flow through a shower. A family of 5 people showers for an average of 6 minutes per person every morning. How many gallons of water does the family use for showering every morning?

A. 75 gallons

B. 60 gallons

C. 65 gallons

D. 70 gallons

Look at the following sale receipt:

8TERS RESTAURANT

Fast Food

143 doty circle, West
Springfield, MA 01089 ,USA
(1234) 254 545

TABLE - 7

Qty	Item	Price
2	Burger	$48
3	Soda	$29
1	Beer	$14
1	Pizza Slice	$12

Visa Sale

Subtotal	$?
Tax	$2.06

Total: $?

7. How many items were purchased?

A. 6

B. 7

C. 8

D. 4

8. Which item was the most expensive?

A. Soda

B. Burger

C. Bear

D. Pizza slice

9. What is the cost of each burger?

A. $48

B. $12

C. $29

D. $24

10. What is the subtotal amount?

A. $105

B. $110

C. $103

D. $100

11. What is the total amount?

A. $105.06

B. $103.06

C. $100.94

D. $106.25

In the cryptocurrency market,
1 Ethereum = $3,320.60.
What is the value of $500,000 in Ethereum? (Round your answer to 2 decimals.)

A. 0.06 ETH

B. 145.50 ETH

C. 150.58 ETH

D. 1,505.80 ETH

13. Bobby's pool can hold a maximum of 3,120 gallons of water. The pool already contains 1,200 gallons of water. Bobby begins to add more water at a rate of 24 gallons per minute. How many minutes can Bobby continue to add water without exceeding the maximum number of gallons?

A. 90 minutes

B. 80 minutes

C. 180 minutes

D. 120 minutes

14. Solve the following inequality:

$$0.05p + 8 \leq 10$$

A. $p \geq 40$

B. $p \leq 40$

C. $p \geq 20$

D. $p \leq 10$

15. Twelve cups of coffee and 8 bagels cost \$66. Ten cups of coffee and 20 bagels cost \$95. What is the cost of each item?

A. Coffee = \$3.50, Bagel = \$3.00

B. Coffee = \$3.00, Bagel = \$3.50

C. Coffee = \$2.50, Bagel = \$3.50

D. Coffee = \$2.00, Bagel = \$3.00

16. Solve the following simultaneous linear equations:

$$2m + 3n = 37$$
$$m - n = 1$$

A. m = 7 and n = 8

B. m = 8 and n = 7

C. m = 7 and n = 7

D. m = 6 and n = 8

17. Let T(x) be the outside temperature (°F) x hours after 8 AM. What does the statement T(2) = 80 mean?

A. The outside temperature is 2°F, 80 hours after 8 AM.

B. The outside temperature is 80°F at 2 PM.

C. The outside temperature is 80°F at 10 AM.

D. After 80 hours, the outside temperature is 2°F.

18. Given $(x) = \frac{0.5x+1.25}{2}$, what is f(0.5) ?

A. 0

B. 0.50

C. 0.25

D. 0.75

19. The formula for converting degrees Celsius to degrees Kelvin is $F = 1.8K - 459.67$, where F is the temperature in degrees Fahrenheit and K is the temperature in degrees Kelvin. If the temperature is 264 degrees Fahrenheit, what is it in Kelvin?

A. 15.53 °K

B. 402.04 °K

C. 390.17 °K

D. 405 °K

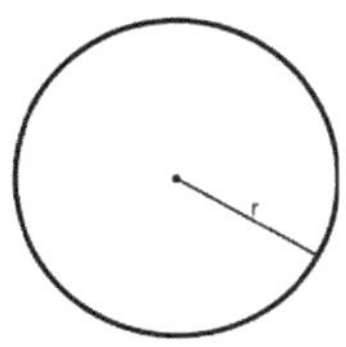

The formula for the area of the circle is

$$A = \pi r^2$$

A is the area of the circle and r is its radius.

20. What is the formula for the radius of the circle?

A. $r = \frac{2A}{\pi}$

B. $r = \frac{A}{2\pi}$

C. $r = \sqrt{\frac{A}{\pi}}$

D. $r = \sqrt{\frac{\pi}{A}}$

21. What is the volume of a sphere with a radius of 10 inches? (Use $\pi = 3.14$)

A. 4,186.67 in^3

B. 1,256.64 in^3

C. 418.88 in^3

D. 5,123,86 in^3

22. If the volume of a cube is 512 cubic inches, what is the length of each side?

A. 9 inches

B. 17 inches

C. 22 inches

D. 8 inches

23. What is the surface area of the following shape?

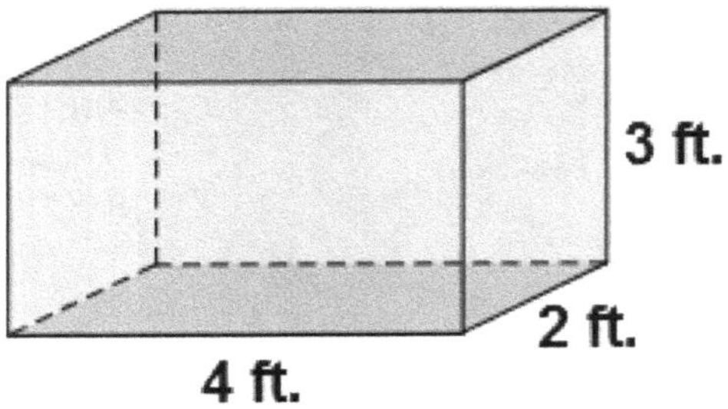

A. 26 ft^2

B. 52 ft^2

C. 24 ft^2

D. 36 ft^2

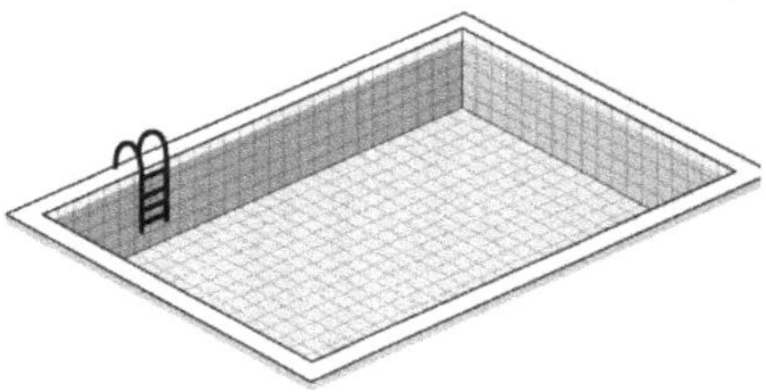

A swimming pool is 25 feet long, 18 feet wide and 7 feet deep. The water-resistant paint needed for the pool costs 1.50 dollars per square feet.

24. What is the volume of the pool?

A. 1,502 cubic feet

B. 3,260 cubic feet

C. 3,150 cubic feet

D. 3,500 cubic feet

25. How much will it cost to paint the interior surfaces of the pool?

A. $1,052

B. $2,253

C. $3,150

D. $1,578

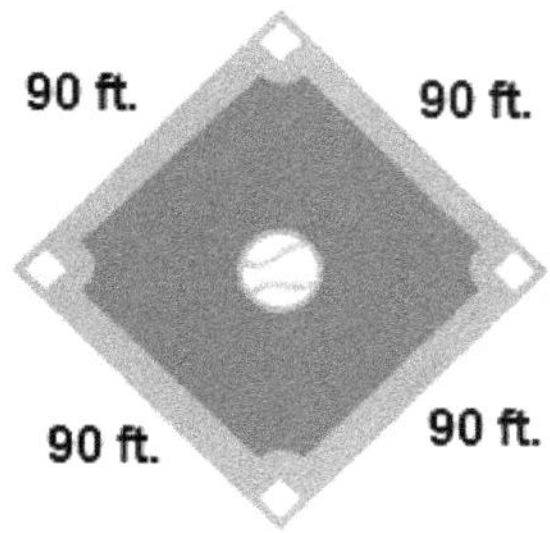

A baseball diamond is a square that is 90 feet on each side. What is the distance a player has to throw the ball from first base to third base?

A. 180 ft.

B. 130.50 ft.

C. 127.28 ft.

D. 189.73 ft.

27. Luke leaves the house to go to school. He walks 300 feet east and 400 feet north. How far away is he from his starting point?

A. 700 feet

B. 500 feet

C. 600 feet

D. 550 feet

The following right triangles are similar:

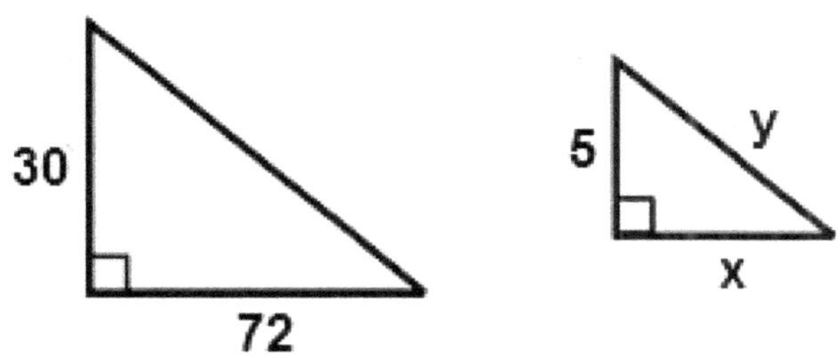

28. What is x?

A. 12

B. 13

C. 24

D. 18

29. What is y?

A. 18

B. 16

C. 13

D. 15

30. Which of the following shapes are congruent?

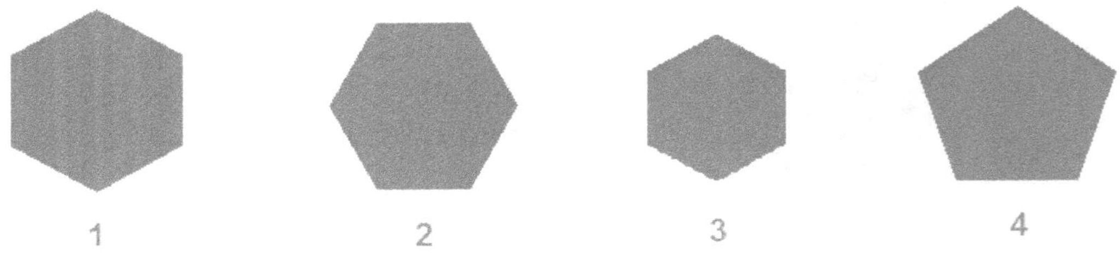

A. Shape 1 and Shape 3

B. Shape 4 and Shape 2

C. Shape 2 and Shape 1

D. All shapes are congruent

31. The population density of a town is 112 people per square mile. 10,640 people are living in the town. How many square miles is the town?

A. 90 square miles

B. 110 square miles

C. 11 square miles

D. 95 square miles

32. There are 15 red cards and 25 blue cards in a box. If one card is chosen randomly, what is the probability of getting a red card?

A. 37.50%

B. 45%

C. 60%

D. 35.50%

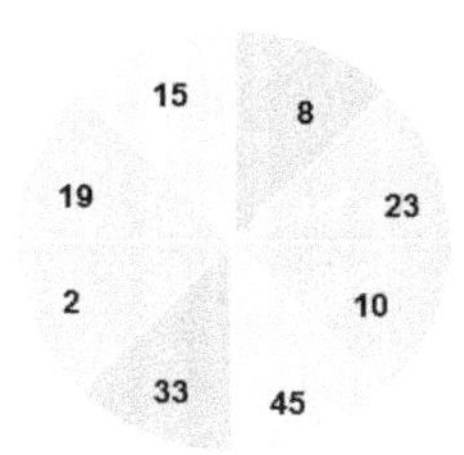

A spinner wheel has 8 equal sectors colored with some numbers.

33. What is the probability of the pointer landing on an even number after spinning the spinner?

A. 0.25

B. 0.375

C. 0.50

D. 0.125

34. What is the probability of the pointer landing on a number greater than 30 after spinning the spinner?

A. 0.75

B. 0.20

C. 0.25

D. 0.45

35. What is the probability of landing on a yellow sector after spinning the spinner?

A. 0.5

B. 0.1

C. 1

D. 0

The following two-way table shows data on nationality and eye color.

	Brown or Hazel Eyes	Blue or Green Eyes	Total
English	83	29	112
Italian	21	13	34
German	65	A	B
Total	169	84	253

36. What is A?

A. 42

B. 56

C. 39

D. 52

37. What is B?

A. 110

B. 107

C. 98

D. 105

38. Are Italians more likely to have blue or green eyes than Germans?

A. Yes

B. No

C. Cannot be determined

39. The grades obtained by a student in five math tests are 77, 65, 81, X, and 72. If the mean of the grades is 75, what is X?

A. 77

B. 80

C. 82

D. 90

40. What is the mode of the following data set: 0.05, 0.011, 0.04, 0.05, 0.031, 0.04, 0.303, 0.111, 0.05, and 0.005?

A. 0.011

B. 0.04

C. 0.111

D. 0.05

REFLECTION ON LEARNING

Answer the following reflection questions and discuss your responses with your teacher or a classmate.

1- How do you feel about your performance on the test?

2- Which types of questions were difficult for you?

3- How do you feel about your time management strategies?

4- What specific things do you want to do differently next time? List them.

5- What math functions or content areas do you want to review? List them.

6- What else do you want your teacher to know?

Answer Keys:

1) D	11) A	21) A	31) D
2) B	12) C	22) D	32) A
3) C	13) B	23) B	33) B
4) C	14) B	24) C	34) C
5) D	15) A	25) B	35) D
6) A	16) B	26) C	36) A
7) B	17) C	27) B	37) B
8) B	18) D	28) A	38) B
9) D	19) B	29) C	39) B
10) C	20) C	30) C	40) D

PRACTICE TEST 2

Read each question and choose the best answer. You have 90 minutes to answer 40 questions.

1. A group of 48 students writes essays in a class. Each student writes 5 essays, and each essay contains an average of 1,200 words. How many words do the students write?

   A. 288,000 words

   B. 57,600 words

   C. 144,000 words

   D. 11,520 words

2. What is m in the following proportion?

$$\frac{1.6\ pounds}{2.7\ inches} = \frac{2.4\ pounds}{m}$$

   A. 5.4 inches

   B. 4.5 inches

   C. 4.05 inches

   D. 1.50 inches

3. John can drive 375 miles on a tank of 20 gallons. How far can it drive on 50 gallons?

   A. 800 miles

   B. 150 miles

   C. 937.5 miles

   D. 550 miles

4. An ounce of sugar has about 110 calories. If a handful of sugar has 415 calories, what's the weight of the sugar in ounces?

   A. 0.27 ounces

   B. 45.65 ounces

   C. 5.34 ounces

   D. 3.77 ounces

5. If a = 1 and b = 2, which of the following is an irrational number?

   A. $\sqrt{b+2}$

   B. $\frac{a}{b}$

   C. $\sqrt{a}$

   D. $\sqrt{a+b}$

6. Which of the following is true?

   A. All radicals are irrational.

   B. All fractions are irrational.

   C. $\sqrt{7}$ is a radical.

   D. $\sqrt{4.5}$ is not irrational.

7. In the news, you hear, “Tuition is expected to increase by 5% next year.” If tuition this year was $1,350 per quarter, what would it be next year?

A. $1,417.50
B. $2,025
C. $1,500.25
D. $2,150

At a supermarket, Cookies Delights costs $18 for a 10-pound bag, Nature’s Cookies costs $20 for a 12-pound bag, and Smart Cookies costs $10 a 5-pound bag.

8. What is the unit price for Nature´s Cookies?

A. $0.60 per pound
B. $1.67 per pound
C. $1.80 per pound
D. $0.56 per pound

9. What is the cost of a 6-pound bag of Smart Cookies?

A. $15
B. $13
C. $12
D. $18

10. What is the better deal?

A. Nature’s Cookies
B. Smart Cookies
C. Delights Cookies

11. Jessica purchased a new laptop for $950. If sales tax is 6.5%, what was the total of her purchase?

A. $1,007
B. $1,011.75
C. $1,050
D. $888.25

12. A pair of shoes originally cost $92, but there’s a 15% discount on them. If the sales tax rate is 5% on the discounted price, how much would you pay in total, including tax, after the discount?

A. $83
B. $82.80
C. $81
D. $82.11

13. Solve the following inequality:

$$8y - 5 > 3y$$

A. $y > 0.5$
B. $y < 1$
C. $y > 1$
D. $y > 0.45$

14. There are 45 animals in a farm. Some are chickens and some are pigs. There are 130 legs in all. How many of each animal are there?

    A. 25 pigs and 20 chickens

    B. 20 pigs and 25 chickens

    C. 15 pigs and 30 chickens

    D. 12 pigs and 33 chickens

15. Solve the following the simultaneous linear equations.

$$p + q = 150$$
$$p - 2q = 0$$

    A. p = 75 and q = 75

    B. p = 50 and q = 100

    C. p = 100 and q = 50

    D. p = 70 and q = 80

16. It costs the class $30 to make cookies for the bake sale. How many cookies must they sell at 20 cents each to make a profit?

    A. Fewer than 100 cookies

    B. More than 150 cookies

    C. Fewer than 150 cookies

    D. At least 80 cookies

17. Given $h(t) = 2t^2 + 5t + 10$, what is $h(0) + h(1)$?

    A. 30

    B. 20

    C. 25

    D. 27

18. The function A(m) = 30m + 700 represents the amount of money denoted as A, in a savings account as a function of time in months, denoted as m. What does the statement A(3) mean?

    A. The amount of money in the savings account after 3 months

    B. The number of months

    C. The amount of money in the savings account 3 months ago

    D. Three times the amount of money in the savings account

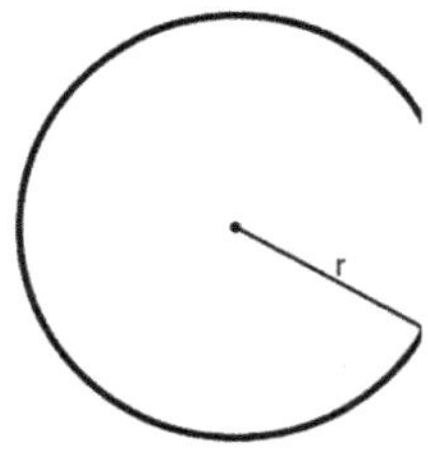

The formula for the circumference of the circle is

$$C = 2\pi r$$

Where C is the circumference of the circle and r is its radius.

19. What is the formula for the radius of the circumference?

A. $r = 2\pi C$

B. $r = \frac{2\pi}{C}$

C. $r = \frac{C}{2\pi}$

D. $r = \frac{\pi}{2C}$

20. What is the circumference of a circle with a radius of 8 inches? (Use $\pi = 3.14$)

A. 25.12 inches

B. 50.24 inches

C. 24.14 inches

D. 55.60 inches

21. If the circumference of a circle is 24 π feet, what is the radius of the circumference?

A. 24 feet

B. 18 feet

C. 48 feet

D. 12 feet

22. A cylindrical water tank has a radius of 6 feet and a height of 15 feet. What is the maximum volume that the water tank can hold? (Use $\pi = 3.14$)

A. 791.30 cubic feet

B. 282.74 cubic feet

C. 1,695.60 cubic feet

D. 1,500 cubic feet

23. A cake decorator rolls a piece of stiff paper to form a cone. He cuts off the tip of the cone and uses it as a funnel to pour decorative sprinkles into containers. The cone has a radius of 4 inches and a height of 10 inches. What is the volume of the cone before the end is cut off? (Use $\pi = 3.14$)

A. 185.52 cubic inches

B. 502.41 cubic inches

C. 135.28 cubic inches

D. 167.47 cubic inches

24. What is the surface area of a cube whose volume is 0.125 cubic feet?

A. 1 ft^2

B. 0.5 ft^2

C. 1.5 ft^2

D. 3 ft^2

Electronics 3000 is selling 50 - inch Smart TV's. This measurement is **the diagonal** distance across the screen. If the screen measures 44 inches in width, what is the actual height of the screen?

A. 66.60 inches

B. 23.75 inches

C. 25 inches

D. 21inches

26. David is building a slide for his kids. The ladder is 10 feet tall and the slide is 12.5 feet long. What is the distance between the base of the ladder and the bottom of the slide?

A. 7.5 feet

B. 16 feet

C. 12 feet

D. 9.5 feet

27. The following rectangles are similar. What is x?

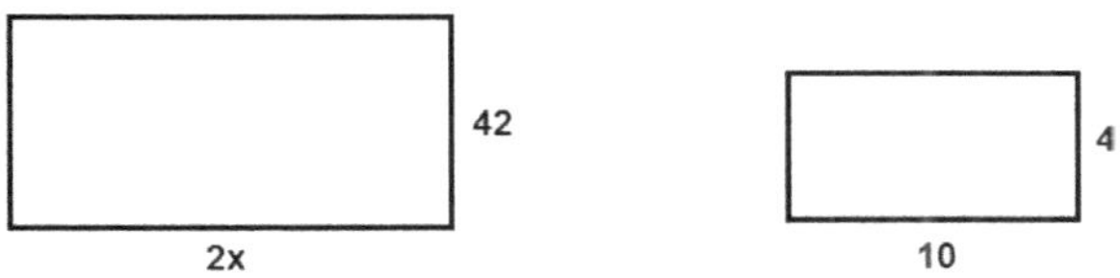

A. 105

B. 64

C. 56

D. 52.5

Triangles 1 and 2 are similar.

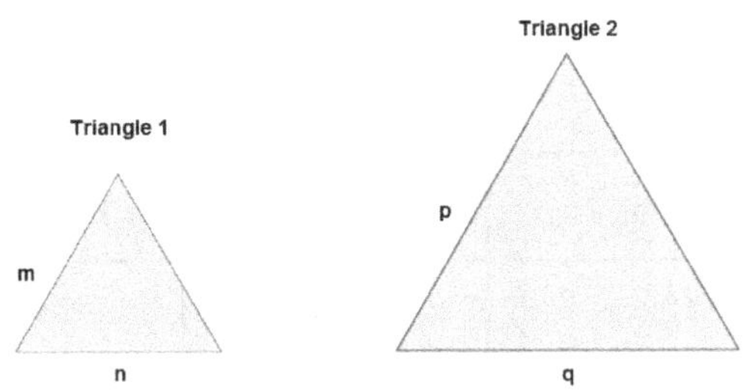

28. Which of the following is true?

A. $m = p$

B. $\frac{m}{q} = \frac{n}{q}$

C. $mn = pq$

D. $\frac{m}{p} = \frac{n}{q}$

29. If m = 2, n = 3, and p = 26, what is q?

A. 36

B. 39

C. 13

D. 35

30. Suppose that Triangle 1 and Triangle 2 are congruent. If the area of Triangle 2 is 50 square inches, what is the area of Triangle 1?

A. 25 square inches

B. 40 square inches

C. 50 square inches

D. Cannot be determined

31. Which of these numbers cannot be a probability?

A. 100%

B. 0.007

C. 3/31

D. 1.13

32. Julia writes each letter of the alphabet on a different slip of paper and puts the slips into a hat. What is the probability of drawing one slip of paper from the hat at random and getting a consonant?

A. 5/26

B. 21/26

C. 2/13

D. 7/13

33. What is the probability of getting a number greater than 4 when a die is rolled?

A. ½

B. 1/6

C. 2/5

D. 1/3

The following two-way table shows the preference for cats or dogs in a high school:

Preference	Male	Female
Cats	58	113
Dogs	x	47
No preference	12	y
Total	193	179

34. What is x?

A. 135

C. 123

B. 113

D. 107

35. How many students prefer dogs?

A. 47

C. 160

B. 170

D. 155

36. What is y?

A. 21

C. 18

B. 20

D. 19

The following box plot summarizes the data for the average monthly high temperatures in degrees Fahrenheit for a city.

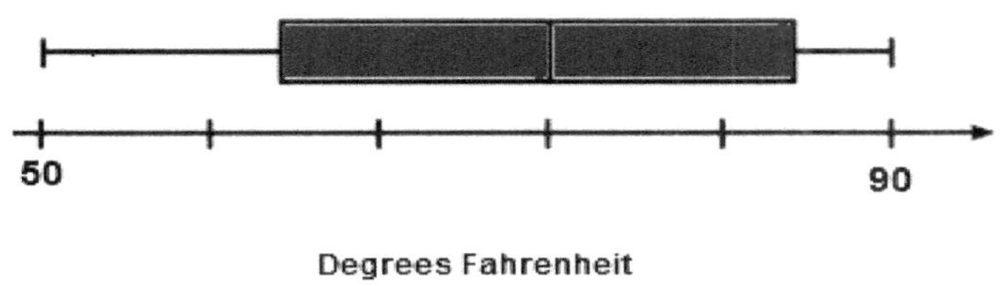

37. What is the maximum temperature registered?

A. 50°F

C. 70°F

B. 90°F

D. 60°F

38. What is the median of the data set?

A. 66°F

C. 72°F

B. 82°F

D. 74°F

39. What is the range of the data set?

A. 40°F

C. 50°F

B. 60°F

D. 90°F

40. What is the value that occurs most often in a given data set?

A. Mean

C. Mode

E. Outlier

B. 0

D. Median

REFLECTION ON LEARNING

Answer the following reflection questions and discuss your responses with your teacher or a classmate.

1- How do you feel about your performance on the test?

2- What types of questions were difficult for you?

3- How do you feel about your time management strategies?

4- What specific things do you want to do differently next time? List them.

5- What math functions or content areas do you want to review? List them.

6- What else do you want your teacher to know?

Answer Keys:

1) A	11) B	21) D	31) D
2) C	12) D	22) C	32) B
3) C	13) C	23) D	33) D
4) D	14) B	24) C	34) C
5) D	15) C	25) B	35) B
6) C	16) B	26) A	36) D
7) A	17) D	27) D	37) B
8) B	18) A	28) D	38) D
9) C	19) C	29) B	39) A
10) A	20) B	30) C	40) C

About CBL

At CBL, we promote systematic solutions, learner-centered textbooks, and forward-thinking strategies in adult education, workforce development, and vocational training. Our diverse solutions and products are intricately designed to enrich students' learning experiences while making the job of busy, hard-working adult instructors easier.

CBL takes pride in publishing student-centered textbooks designed to prepare learners for CASAS, TABE 11&12, HiSET, and GED assessments and assist instructors in covering course curricula and standards with confidence.

Our publications also include teaching guides, test prep tools, and study guides that foster reflective learning, ensuring sustained engagement in active learning. Find our meticulously crafted textbooks on our book page (cbledu.com) or major platforms like Amazon, Barnes & Noble, and Ingram Spark.

CBL also guides adult education and workforce programs in establishing robust professional development programs—training, peer-mentoring, coaching, community of practices (CoPs), and instructional systems— fostering a culture of continuous improvement and contributing to higher learner retention and success rates. We also offer workshops and PD sessions for adult educators and classroom instructors.

If you have questions about instructional systems, textbooks, or student learning and retention, contact us today at teamcbl@cbledu.com or 410-960-4082.

CBL | COACHING FOR BETTER LEARNING

www.ingramcontent.com/pod-product-compliance
Lightning Source LLC
LaVergne TN
LVHW082355200926
848481LV00012B/3574